Lecture Notes in Computer Science   11517

Commenced Publication in 1973
Founding and Former Series Editors:
Gerhard Goos, Juris Hartmanis, and Jan van Leeuwen

More information about this series at http://www.springer.com/series/7408

Yunni Xia · Liang-Jie Zhang (Eds.)

Services – SERVICES 2019

15th World Congress
Held as Part of the Services Conference Federation, SCF 2019
San Diego, CA, USA, June 25–30, 2019
Proceedings

 Springer

Editors
Yunni Xia
Chongqing University
Chongqing, China

Liang-Jie Zhang ⓘ
Kingdee International Software
Group Co., Ltd.
Shenzhen, China

ISSN 0302-9743   ISSN 1611-3349 (electronic)
Lecture Notes in Computer Science
ISBN 978-3-030-23380-8   ISBN 978-3-030-23381-5 (eBook)
https://doi.org/10.1007/978-3-030-23381-5

LNCS Sublibrary: SL2 – Programming and Software Engineering

This Springer imprint is published by the registered company Springer Nature Switzerland AG
The registered company address is: Gewerbestrasse 11, 6330 Cham, Switzerland

Preface

The aim of the 2019 World Congress on Services (SERVICES 2019) was to provide an international forum to attract researchers, practitioners, and industry business leaders in all services sectors to help define and shape the modernization strategy and directions of the services industry.

SERVICES 2019 was part of the Services Conference Federation (SCF). SCF 2019 had the following ten collocated service-oriented sister conferences: 2019 International Conference on Web Services (ICWS 2019), 2019 International Conference on Cloud Computing (CLOUD 2019), 2019 International Conference on Services Computing (SCC 2019), 2019 International Congress on Big Data (BigData 2019), 2019 International Conference on AI & Mobile Services (AIMS 2019), 2019 World Congress on Services (SERVICES 2019), 2019 International Congress on Internet of Things (ICIOT 2019), 2019 International Conference on Cognitive Computing (ICCC 2019), 2019 International Conference on Edge Computing (EDGE 2019), and 2019 International Conference on Blockchain (ICBC 2019). As the founding member of SCF, the First International Conference on Web Services (ICWS) was held in June 2003 in Las Vegas, USA. The First International Conference on Web Services—Europe 2003 (ICWS-Europe 2003) was held in Germany in October 2003. ICWS-Europe 2003 was an extended event of the 2003 International Conference on Web Services (ICWS 2003) in Europe. In 2004, ICWS-Europe was changed to the European Conference on Web Services (ECOWS), which was held in Erfurt, Germany. To celebrate its 16th birthday, SCF 2018 was held successfully in Seattle, USA.

This volume presents the accepted papers for the 2019 World Congress on Services (SERVICES 2019), held in San Diego, USA, during June 25–30, 2019. All topics regard software engineering foundations and applications, with a focus on novel approaches for engineering requirements, design and architectures, testing, maintenance and evolution, model-driven development, software processes, metrics, quality assurance and new software economics models, search-based software engineering, benefiting day-to-day services sectors and derived through experiences, with a focus on scale, pragmatism, transparency, compliance, and/or dependability.

We accepted 11 papers, including nine full papers and two short papers. Each was reviewed and selected by at least three independent members of the SERVICES 2019 international Program Committee. We are pleased to thank the authors, whose submissions and participation made this conference possible. We also want to express our thanks to the Organizing Committee and Program Committee members, for their dedication in helping to organize the conference and in reviewing the submissions. We look forward to your great contributions as a volunteer, author, and conference participant for the fast-growing worldwide services innovations community.

May 2019
Yunni Xia
Liang-Jie Zhang

Organization

Program Chair

Yunni Xia   Chongqing University, China

Services Conference Federation (SCF 2019)

SCF 2019 General Chairs

Calton Pu   Georgia Tech, USA
Wu Chou   Essenlix Corporation, USA
Ali Arsanjani   8x8 Cloud Communications, USA

SCF 2019 Program Chair

Liang-Jie Zhang   Kingdee International Software Group Co., Ltd., China

SCF 2019 Finance Chair

Min Luo   Services Society, USA

SCF 2019 Industry Exhibit and International Affairs Chair

Zhixiong Chen   Mercy College, USA

SCF 2019 Operations Committee

Huan Chen   Kingdee International Software Group Co., Ltd., China
Jing Zeng   Kingdee International Software Group Co., Ltd., China
Liping Deng   Kingdee International Software Group Co., Ltd., China
Yishuang Ning   Tsinghua University, China
Sheng He   Tsinghua University, China

SCF 2019 Steering Committee

Calton Pu (Co-chair)   Georgia Tech, USA
Liang-Jie Zhang (Co-chair)   Kingdee International Software Group Co., Ltd., China

Services 2019 Program Committee

Hong Xie   The Chinese University of Hong Kong, SAR China
Xiuhua Li   University of British Columbia, Canada
Zhaohong Jia   Anhui University, China
Xin Luo   Chongqing Institute of Green and Intelligent
                                 Technology, Chinese Academy of Sciences, China

Zhiming Zhao   University of Amsterdam, The Netherlands
Stefano Sebastio   Inria Rennes, Bretagne Atlantique, France
Zheng Zheng   Beihang University, China

Contents

SMT-Based Modeling and Verification
of Cloud Applications

Xiyue Zhang and Meng Sun[✉]

Department of Informatics and LMAM, School of Mathematical Sciences,
Peking University, Beijing, China
{zhangxiyue,sunm}@pku.edu.cn

Abstract. Cloud applications have been rapidly evolving and gained
more and more attention in the past decade. Formal modeling and verification of cloud services are necessarily needed to guarantee their correctness and reliability of complex cloud applications. In this paper, we
present a formal framework for modeling and verification of cloud applications based on the SMT solver Z3. Simple cloud services are specified
as the basis for the modeling of composition and more complex cloud
services. Three different classes *Service*, *Composition* and *Cloud* indicating simple cloud services, composition patterns and composed cloud services are defined, which facilitates the further development of attributes
and methods. We also propose an approach to check the refinement and
equivalence relations between cloud services, in which counter examples
can be automatically generated when the relation is not valid.

Keywords: Cloud applications · Services · Z3 · Modeling · Verification

1 Introduction

With the rapid development of big data and cloud computing technologies, cloud
applications spring up quickly in the past decade. The increasing complexity of
modern cloud applications has changed the perspective of software designers who
now have to consider large-scale applications consisting of a colossal number of
services and featuring complex interaction mechanisms. Nowadays, cloud applications are usually distributed, heterogeneous, decentralized and safety-critical,
having complex concurrent behavior and are operating in unpredictable environments, and it is notoriously difficult to guarantee their trustworthy. Therefore,
formal verification of cloud applications becomes the focus of attention in both
academic and industry.

Although current researches on cloud computing are mostly focused on technical problems such as resource allocation [8] and task scheduling [17], there are
some attempts to the formalization of fundamental notions in cloud computing. For example, an abstract formal model of cloud workflows was proposed
in [7] using the Z notation. In [6], the hierarchical colored Petri Net model was
adopted to specify the security mechanism in cloud computing. The Petri Net

© Springer Nature Switzerland AG 2019
Y. Xia and L.-J. Zhang (Eds.): SERVICES 2019, LNCS 11517, pp. 1–15, 2019.
https://doi.org/10.1007/978-3-030-23381-5_1

model is also used in [4] to build the fault tolerant model of cloud computing, and as the basis for a dynamic fault tolerant strategy in cloud computing. The agent paradigm was adopted in [14] to manage cloud resources and support cloud service discovery, negotiation and composition. A Bigraph model was proposed in [3] to formally specify cloud services and customers and their interaction schemes. In [11], Event-B is integrated with discrete-event simulation to analyze the performance and reliability of resilience of data stored in the cloud.

To support rigorous development of cloud applications and enhance their trustworthy, only providing the formal specification is certainly not enough and we need to further investigate the formal verification techniques that help for understanding and reasoning about cloud applications and ensuring their trustworthy. Model checking and theorem proving are two most-widely used verification techniques. In [1], the model checker UPPAAL is used to synthesize an optimal infinite scheduler for a given specification of Mobile Cloud Computing systems. In [13] a holistic approach was proposed to verify the correctness of Hadoop cloud architectures using model checking techniques. Although the model checking approach is powerful and fully automatic, the state-space explosion is an inherent problem for all the model checking approaches which is serious for large-scale systems like cloud applications. On the other hand, theorem proving technique is used in [12] to verify properties of cloud services and their compositions in PVS, which is based on the relational UTP (Unifying Theories of Programming) [9] semantics for cloud services that has been proposed in [16]. However, automatic verification in interactive theorem provers like PVS is a hard problem due to the undecidable algorithms and proof methods. Furthermore, sometimes users may fail to prove that a property holds or not, and can not produce a counter example for it using theorem provers.

SMT-based techniques have been used extensively for program verification [2]. In this paper, we propose a formal framework for verification of cloud applications using the SMT solver Z3 [5], which is also based on the relational UTP semantics for cloud services [16]. UTP aims to formalize the similar features of different languages in a similar style. It has been proved to be appropriate for formal semantics of various programming languages and specification languages like rCOS [10] and Reo [15]. Z3 is a state-of-the-art SMT solver, which can be used to check the satisfiability of logical formulas over one or more theories. It provides bindings for various programming languages. In this paper, we use Z3 python-bindings to specify the models and develop the verification framework for cloud applications. Unlike other theorem provers, such as PVS, Coq, etc., Z3 can automatically generate counter examples or prove the validity of a specific goal.

The paper is organized as follows: Sect. 2 presents how simple cloud services/applications are specified in Z3 based on observations on their input and output ports. The models of a family of composition patterns, which are used to construct more complex cloud services are presented in Sect. 3. In Sect. 4, we use the refinement-relation-check function for complex cloud services as an example to show how to verify properties of cloud services and applications, and provide some case studies. Finally, Sect. 5 summarizes the paper.

2 Formalization of Cloud Services in Z3

Usually the computing services in a cloud application are distributed over the internet and far from its clients. Clients have no knowledge about the implementation details of the services and configuration of the application, and can access the cloud application regardless of their locations or what device being used. The only possible way that clients can know about a cloud application is via observations on the services provided by the application at corresponding input/output ports.

2.1 Observations as Timed Data Streams

A cloud service $\mathbf{C}$ is interpreted as a relation between an initial observation on inputs to $\mathbf{C}$ and a subsequent observation of the behavior of $\mathbf{C}$. we use $in_{\mathbf{C}}$ and $out_{\mathbf{C}}$ to denote what happen as inputs and outputs of a cloud service $\mathbf{C}$, respectively.

For every port of a cloud service $\mathbf{C}$, the corresponding observation on it is given by a *timed data stream* (TDS), which is defined as follows:

Definition 1. *Let D be a set of data elements and $\mathbb{R}_+$ be the set of non-negative real numbers which is used to represent time moments. Let $DS = D^{\omega}$ be the set of data streams, that is, the set of all streams $\alpha = (\alpha(0), \alpha(1), \alpha(2), \ldots)$ over D, and $\mathbb{R}_+^{\omega}$ be the set of all streams $a = (a(0), a(1), a(2), \ldots)$ over $\mathbb{R}_+$. The set of time streams is defined by the following subset of $\mathbb{R}_+^{\omega}$:*

$$TS = \{a \in \mathbb{R}_+^{\omega} \,|\, \forall n \geq 0.a(n) < a(n+1) \wedge \forall t \in \mathbb{R}_+.\exists k \in \mathbb{N}.a(k) > t\}$$

For two time streams a and b, $a < b \equiv \forall n \geq 0.a(n) < b(n)$. A timed data stream is defined as a pair $\langle \alpha, a \rangle$ consisting of a data stream $\alpha \in DS$ and a time stream $a \in TS$. We use TDS to denote the set of timed data streams.

Let $I_{\mathbf{C}}$ and $O_{\mathbf{C}}$ be the set of input and output port names of $\mathbf{C}$, the specification focuses on the constraints on the timed data streams of the corresponding ports:

$$in_{\mathbf{C}} : I_{\mathbf{C}} \to TDS$$

$$out_{\mathbf{C}} : O_{\mathbf{C}} \to TDS$$

We use relations on timed data streams to model cloud services. Every cloud service $\mathbf{C}$ can be represented by a pair of predicates P and Q as follows:

$$\mathbf{C}(in : in_{\mathbf{C}}; out : out_{\mathbf{C}})$$

$$\mathbf{pre} : P(in_{\mathbf{C}})$$

$$\mathbf{post} : Q(in_{\mathbf{C}}, out_{\mathbf{C}})$$

where $\mathbf{C}$ is the name of the cloud service, $P(in_{\mathbf{C}})$ is the condition that should be satisfied by inputs $in_{\mathbf{C}}$ of the cloud service, and $Q(in_{\mathbf{C}}, out_{\mathbf{C}})$ is the condition that should be satisfied by outputs $out_{\mathbf{C}}$ of $\mathbf{C}$.

2.2 Modeling of Simple Cloud Services

We first present the formal definitions of simple cloud services, which are used as the basis for the specification of compositions and development of more complex cloud services. The simple cloud services are dealt with as object instances of *Class* Service. An object instance is attached with four attributes: *service type, pre-condition, nodes* (represent timed data streams), and its corresponding *method*. We specify the time and data constraints of relations between inputs and outputs in the service definitions (methods). Basically, the behavior constraints of cloud services consist of two parts: *data constraints* and *time constraints*.

SyncSer has an input port whose corresponding timed data stream is specified as `nodesParam[0]` and a set of output ports (at least one, guaranteed by the assertion of the method). The behavior pattern of this cloud service follows like this: every item in both data stream and time stream at the output ports should be exactly equal to the items in the input data stream and time stream. All the conditions are specified as constraints. Finally, we take the conjunction of every constraint in the constraint list if the pre-condition of the service is True.

```
1     def SyncSer(self, bound, nodesParam):
2         if self.preCondition == True:
3             assert len(self.nodes) >= 2
4             target_num = len(self.nodes) - 1
5             constraints = []
6             for i in range(target_num):
7                 for j in range(bound):
8                     constraints += [ nodesParam[0]['data'][j] ==
                          nodesParam[i + 1]['data'][j] ]
9                     constraints += [ nodesParam[0]['time'][j] ==
                          nodesParam[i + 1]['time'][j] ]
10            return Conjunction(constraints)
11        else:
12            return True
```

BufferSer has one input port and one output port. The time constraints of this service type is more complex than *SyncSer*, while the data-related constraints are the same as *SyncSer*. The time when the current data item gets transferred into the service through the input port is strictly earlier than the time when the data item gets taken/read through the output port. Besides, the time of the next request (input data item) should be strictly later than the previous returned result (last output data item).

```
1     def BufferSer(self, bound, nodesParam):
2         if self.preCondition == True:
3             assert len(self.nodes) == 2
4             constraints = []
5             for i in range(bound):
6                 constraints += [ nodesParam[0]['data'][i] ==
                      nodesParam[1]['data'][i] ]
7                 constraints += [ nodesParam[0]['time'][i] <
                      nodesParam[1]['time'][i] ]
```

```
8           for i in range(bound - 1):
9               constraints += [ nodesParam[0]['time'][i + 1] >
                    nodesParam[1]['time'][i] ]
10          return Conjunction(constraints)
11      else:
12          return True
```

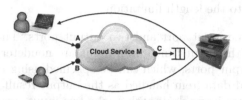

Fig. 1. Remote printing service

Example 1. Consider a simple example where a remote printer offers its printing service to two clients, which compete for the use of this shared resource. Each client can send out multiple printing requests to the printer and the requests from different clients are placed in a queue to be processed by the printer in a first-come first-served manner. After a file is printed out it can be collected by the client later. In order to keep the example simple to expose without considering priority for scheduling different printing tasks, we assume that requests from different clients never arrive simultaneously.

The cloud service **M** in Fig. 1 receives requests from different clients at ports A and B, and delivers a sequence of requests through port C to a queue on the printer side. Such a service is called *MergeSer*. It has two input ports and one output port. When there is only one timed data stream from one specifc input port, the *MergeSer* is reduced to the *SyncSer* with one input and one output ports. When there exists two different timed data streams from the input ports, the *MergeSer* is captured through the method *Merge*, which is defined recursively. Every time one of the timed data items from these two input ports is chosen to be fed into the service. The specification of this service is in the following:

```
1       def MergeSer(self, bound, nodesParam):
2           if self.preCondition == True:
3               assert 2 <= len(self.nodes) <=3
4               if len(self.nodes) == 2:
5                   constraints = []
6                   for i in range(bound):
7                       constraints += [ nodesParam[0]['data'][i] ==
                            nodesParam[1]['data'][i] ]
8                       constraints += [ nodesParam[0]['time'][i] ==
                            nodesParam[1]['time'][i] ]
9                   return Conjunction(constraints)
```

```
10              elif len(self.nodes) == 3:
11                  return self.Merge(bound, nodesParam)
12          else:
13              return True
```

where *Merge* is the method to deal with the case when the two input ports both have requests to provide for the *MergeSer*. It captures the behavior of merging two input timed data streams into one output stream and the detailed definition is omitted here due to the length limitation.

RouterSer has one input port and two output ports. The behavior of this service type is that the input timed data items are nondeterministically taken through the two output ports, which can be specified using the *Merge* method with the input timed data item handled as the output result in *Merge* and the two output timed data items dealt with as the two input requests in *Merge*.

```
1       def RouterSer(self, bound, nodesParam):
2           if self.preCondition == True:
3               assert len(self.nodes) == 3
4               new_nodes =
5               [nodesParam[1], nodesParam[2], nodesParam[0]]
6               return self.Merge(bound, new_nodes)
7           else:
8               return True
```

3 Composition of Cloud Services

Different cloud services can be composed together to build more complex services/applications. Simple cloud services are defined as object instances of class *Service*, therefore their composition can be naturally modeled by composition on *Service* instances, which leads to a new class *Composition* capturing the behavior of the composed cloud service/application. In this section, we introduce a family of composition operations for two cloud services $\mathbf{serv}_i (i = 1, 2)$:

Sequential Composition. Suppose one output port O of $\mathbf{serv}_1$ and one input port I of $\mathbf{serv}_2$ can be joined together and the timed data stream that happens on O thus can be taken as the input on I for $\mathbf{serv}_2$. After joining these two ports, the extra constraints restricted on O (and I) are about the equality between the output timed data stream of $\mathbf{serv}_1$ and the input timed data stream of $\mathbf{serv}_2$. Since there can exist more than one output ports in $\mathbf{serv}_1$, the attribute *index* is designed to indicate the correct output port being joined. The output timed data stream in $\mathbf{serv}_1$ is thus represented by nodesParam1[index] and the input timed data stream in $\mathbf{serv}_2$ is represented by nodesParam2[0]. Then the result specification of the cloud service by sequentially composing $\mathbf{serv}_1$ and $\mathbf{serv}_2$ is:

```
1       def SeqComp(self, bound, nodesParam1, nodesParam2):
2           if self.serv1.preCondition == self.serv2.preCondition ==
                True:
```

```
3          constraints = []
4          for i in range(bound):
5              constraints += [ nodesParam1[self.index]['data'][i]
                   == nodesParam2[0]['data'][i] ]
6              constraints += [ nodesParam1[self.index]['time'][i]
                   == nodesParam2[0]['time'][i] ]
7          return And(
8              self.serv1.valueFunctions[self.serv1.service](bound,
                   nodesParam1),
9              self.serv2.valueFunctions[self.serv2.service](bound,
                   nodesParam2),
10             Conjunction(constraints))
11         else:
12             return True
```

Note that when two cloud services $serv_1$ and $serv_2$ are sequentially composed, we can certainly join more than one pair of ports together and the definition of the resulting service is similar, but it is not necessary to join all the output ports of $serv_1$ to all the input ports of $serv_2$. Some ports in the services can be left as the input/output ports for the resulting service. The definition for the general situation is similar and can be easily obtained.

External, Internal and Conditional Choices. Cloud services can be aggregated in a number of different ways, besides the sequential composition. In the following we consider a few such combinators. A typical composition pattern being widely used is *external choice*. For the two cloud services $serv_1$ and $serv_2$, when they are put together and interacting with the environment, clients from the environment are allowed to choose either to input on the input ports of $serv_1$, or on input ports of $serv_2$, which will trigger the corresponding cloud service $serv_1$ or $serv_2$, respectively, and produce the associated output on the corresponding output ports. Formally, the result specification of the cloud service as an external choice of $serv_1$ and $serv_2$ is defined as:

```
1      def ExChoice(self, bound, nodesParam1, nodesParam2):
2          if self.serv1.preCondition == self.serv2.preCondition ==
               True:
3              return And( self.serv1.valueFunctions[self.serv1.service
                   ](bound, nodesParam1),
4              self.serv2.valueFunctions[self.serv2.service](bound,
                   nodesParam2))
5          elif self.serv1.preCondition == True and self.serv2.
               preCondition == False:
6              return self.serv1.valueFunctions[self.serv1.service](
                   bound, nodesParam1)
7          elif self.serv1.preCondition == False and self.serv2.
               preCondition == True:
8              return self.serv2.valueFunctions[self.serv2.service](
                   bound, nodesParam2)
9          else:
10             return True
```

Sometimes it is possible that both cloud services might have input ports in common so that there is no clear prescription as to which route is followed when one of these common ports is chosen. In the implementation, either service can be chosen to be executed. This case is captured by the *internal choice* pattern, which is formally defined as follows:

```
1       def InChoice(self, bound, nodesParam1, nodesParam2):
2           if self.serv1.preCondition == self.serv2.preCondition ==
                True:
3               return Or( self.serv1.valueFunctions[self.ser1.service](
                    bound, nodesParam1),
4               self.serv2.valueFunctions[self.serv2.service](bound,
                    nodesParam2))
5           else:
6               return True
```

Besides the external and internal choices, a further form of choice, the *conditional choice* which is based on the value of a boolean expression, is also needed for combination of cloud services. This case is formally defined by the following definition which means that if the boolean expression is satisfied then the cloud service $serv_1$ is executed, and otherwise, $serv_2$ is executed:

```
1       def ConChoice(self, bound, nodesParam1, nodesParam2):
2           if self.bool_con == True:
3               return self.serv1.valueFunctions[self.serv1.service](
                    bound, nodesParam1)
4           else:
5               return self.serv2.valueFunctions[self.serv2.service](
                    bound, nodesParam2)
```

Parallel Composition. The simplest form of parallel combinator captures the case that both cloud services $serv_1$ and $serv_2$ are invoked and executed in parallel when triggered by a pair of inputs on the corresponding input ports of both $serv_1$ and $serv_2$. Therefore, to make it possible to execute the parallel combination of $serv_1$ and $serv_2$, both pre-conditions of the two services should be satisfied and the execution will lead to the result that the post-conditions of both $serv_1$ and $serv_2$ should be satisfied.

```
1       def ParallelComp(self, bound, nodesParam1, nodesParam2):
2           if self.serv1.preCondition == self.serv2.preCondition ==
                True:
3               return And(
4               self.serv1.valueFunctions[self.serv1.service]
5                       (bound, nodesParam1),
6               self.serv2.valueFunctions[self.serv2.service]
7                       (bound, nodesParam2))
8           else:
9               return True
```

In the parallel composition defined above, when two cloud services are put into parallel, they may evolve completely autonomously, i.e., we have no restriction on the inputs for the two services and they can arrive at any time. Sometimes we may hope to have some inputs for $serv_1$ and $serv_2$ arrive only simultaneously. For simplicity, we assume that the data can only arrive at the input ports I_1 and I_2 simultaneously, where I_1 and I_2 belong to the input ports of $serv_1$ and $serv_2$ respectively. And the data arriving at all the other input ports except I_1 and I_2 are independent. In this case, extra constraints on the input timed data streams of the two services should be captured. Then we have the following specification:

```
1       def SyncParallel(self, bound, nodesParam1, nodesParam2):
2           if self.serv1.preCondition == self.serv2.preCondition ==
            True:
3               constraints = []
4               for i in range(bound):
5                   constraints += [ nodesParam1[0]['data'][i] ==
                        nodesParam2[0]['data'][i] ]
6                   constraints += [ nodesParam1[0]['time'][i] ==
                        nodesParam2[0]['time'][i] ]
7               return And(
8               self.serv1.valueFunctions[self.serv1.service](bound,
                    nodesParam1),
9               self.serv2.valueFunctions[self.serv2.service](bound,
                    nodesParam2),
10              Conjunction(constraints))
11          else:
12              return True
```

In many cases, a family of cloud services may exist and behave in parallel in a pairwise fashion. To model this, the n-ary version of these two parallel combinators are very helpful. The corresponding specification can be easily generalized to the case for composing multiple services.

A similar situation we consider is the case of merging two input ports of cloud services $serv_1$ and $serv_2$ into one port. Let nodesParami for $i = 1, 2$ be the timed data streams on the input port I_i in $serv_i$, respectively. By merging I_1 and I_2 into one port I, when the resulting service receives a request on I, it will behave in a broadcasting way. In other words, the request will be replicated on I and sent to both $serv_1$ and $serv_2$ to trigger their execution simultaneously. This operation can be realized by renaming without defining a specific method.

4   Verification of Cloud Applications in Z3

Based on the specification of simple cloud services and the family of composition patterns, we can develop more complex cloud services for different needs, which are defined as a new class *Cloud*. In this class, we provide two methods *config* and *compose* to develop complex cloud services out of the simple cloud services and the composition patterns. Meanwhile, we encode the refinement and equivalence relations between cloud services as propositions under analysis. Based on the Z3 SMT solver, we defined the main method *Refine* to check the validity of the refinement and equivalence relations or to generate counter examples for the dissatisfaction of relations.

```
1   class Cloud:
2       def __init__(self):
3           self.services = []
4           self.compositions = []
5
6       def config(self, service, pre_condition, *nodes):
7           self.services += [Service(service, pre_condition, nodes)]
8           return self
9
10      def compose(self, composition, serv1, serv2, index, bool_con =
            True):
11          self.compositions +=
12          [Composition(composition, serv1, serv2, index, bool_con)]
13          return self
```

The predicate capturing the refinement relation between two cloud services is an implication statement. Assume we have two complex cloud services $cServ_1$ and $cServ_2$, $cServ_1$ is a refinement of $cServ_2$ (or $cServ_1$ refines $cServ_2$) if and only if the specification of $cServ_2$ is further restricted by $cServ_1$, i.e., $cServ_1 \rightarrow cServ_2$. Furthermore, two cloud services are in equivalence relation, i.e. $cServ_1 \leftrightarrow cServ_2$, if and only if they refine each other. Such proposition is valid if it is always true whatever the assignment of variables is. It is satisfiable if there exists an evidence (an assignment to the variables) under which this proposition evaluates true. If the proposition $cServ_1 \rightarrow cServ_2$ is valid, i.e. always true under any assignment of values, then its negation will not have any witness (any satisfying assignment). In other words, the negation of the implication is unsatisfiable. On the other hand, if a solution is found for the negation of the proposition, then this solution is actually the counter example for the satisfaction of the relation.

Algorithm 1 provides the pseudocode for the definition of the *Refine* method, in which a solver is created and the constraints of the negation of the target implication are added into the solver. If the solver returns unsat (corresponding to True in Algorithm 1), then it acts as a proof for the validity of the refinement relation between the two cloud services under analysis. If the solver returns sat (corresponding to False in Algorithm 1), the model (witness) of the negation is the counter example we need for proving the dissatisfaction of the relation.

In the beginning, the command Solver() is used to create a general purpose Z3 Solver. The dictionary nodes stores the uninterpreted variables, which can be further used to generate node parameters for invocation of simple cloud services. The first double for loop generates the time constraints (for well-definedness) and constraints for each specific service in $cServ_1$ and the next double for loop generates the time constraints and constraints for each specific composition in

$cServ_1$. The procedures up to now are used to handle $cServ_1$. The subsequent procedures are for the refined cloud service $cServ_2$. The generation of time constraints and constraints for each simple cloud service and composition is similar. Besides, the uninterpreted symbols that need to be under universal quantifier are stored in UniVar.

Example 2. Consider the cloud applications shown in Fig. 2, where users can check the information about flights and can order flight tickets. Different companies offer flights to the same location and the ticket availability and price for each flight varies and may change at any time. If a user wants to make a trip between two places and make a query for the flight information, he/she hopes to collect the information for all the available flights at the latest time.

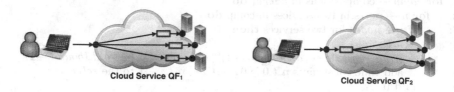

Cloud Service QF₁   Cloud Service QF₂

Fig. 2. Flight query services

The cloud service QF1 in Fig. 2 has a buffer on the server side for each flight company. It accepts queries from users and store queries in every buffer. Based on the simple cloud services, the model of QF1 can be developed in Z3 in the following way:

```
1   QF1 = Cloud()
2   QF1.config('Buffer', 'True', 'A', 'B')
3   QF1.config('Buffer', 'True', 'A', 'C')
4   QF1.config('Buffer', 'True', 'A', 'D')
```

In cloud service QF2, the location of the buffer is on the user's side. QF2 can accept the queries from the user, place the queries in the buffer and simultaneously send a copy of the queries to each flight company. The model of QF2 is constructed like this:

```
1   QF2 = Cloud()
2   QF2.config('Buffer', 'True', 'A', 'E')
3   QF2.config('Sync', 'True', 'E', 'B')
4   QF2.config('Sync', 'True', 'E', 'C')
5   QF2.config('Sync', 'True', 'E', 'D')
```

Next we invoke the *Refine* function in two directions to check the refinement and equivalence relation between QF1 and QF2. The result of QF2.Refine(QF1, 10) is *True* and *None*, which means that QF2 is indeed a refinement of QF1 and therefore no counter example is provided.

Algorithm 1. $cServ_1$.Refine ($cServ_2$, bound)

Require: $cServ_1$ and $cServ_2$ are both instances of cloud services.
Ensure: the function returns *True* or *False with a counter example*
1: $solver \leftarrow$ Solver( ) ; $nodes \leftarrow \{\}$
2: **for** $ser \leftarrow$ services in $cServ_1$ **do**
3:    **for** $n \leftarrow$ ports in ser **do**
4:      **if** $n \notin nodes$ **then**
5:        $nodes[n] \leftarrow$
          $\{time : [n\_t\_0, \cdots, n\_t\_(bound - 1)], \ data : [n\_d\_0, \cdots, n\_d\_(bound - 1)]\}$
6:        add time constraints $n\_t\_0 \geq 0 \wedge n\_t\_i < n\_t\_i + 1$ to the $solver$
7:      **end if**
8:    **end for**
9:    add service constraints to the $solver$ according to definitions in Section 2.2
10: **end for**
11: **for** $comp \leftarrow$ compositions in $cServ_1$ **do**
12:    **for** $n \leftarrow$ ports in two services in comp **do**
13:      **if** $n \notin nodes$ for two services **then**
14:        $nodes[n] \leftarrow$
          $\{time : [n\_t\_0, \cdots, n\_t\_(bound - 1)], \ data : [n\_d\_0, \cdots, n\_d\_(bound - 1)]\}$
15:        add time constraints $n\_t\_0 \geq 0 \wedge n\_t\_i < n\_t\_i + 1$ to the $solver$
16:      **end if**
17:    **end for**
18:    add composition constraints to the $solver$ according to definitions in Section 3
19: **end for**
20: $UniVar \leftarrow \{\}$; $ReSerConstr \leftarrow \{\}$; $RecompConstr \leftarrow \{\}$; $ReConstr \leftarrow \{\}$
21: **for** $ser \leftarrow$ services in $cServ_2$ **do**
22:    **for** $n \leftarrow$ ports in ser **do**
23:      **if** $n \notin nodes$ **then**
24:        generate uninterpreted variables
25:        add the variables to UniVar (variables under the universal quantifier)
26:        add time constraints $n\_t\_0 \geq 0 \wedge n\_t\_i < n\_t\_i + 1$ to ReSerConstr
27:      **end if**
28:      add service constraints to $ReSerConstr$ according to definitions in Section 2.2
29:    **end for**
30: **end for**
31: **for** $comp \leftarrow$ compositions in $cServ_2$ **do**
32:    **for** $n \leftarrow$ ports in two services in comp **do**
33:      **if** $n \notin nodes$ **then**
34:        generate uninterpreted variables
35:        add the variables to UniVar (variables under the universal quantifier)
36:        add time constraints $n\_t\_0 \geq 0 \wedge n\_t\_i < n\_t\_i + 1$ to ReCompConstr
37:      **end if**
38:    **end for**
39:    add composition constraints to $ReCompConstr$ according to definitions in Section 3
40: **end for**
41: $ReConstr = \neg (ReSerConstr \wedge ReCompConstr)$
42: **let** $UniVar$ **be** $\{n_1, \cdots, n_m\}$, add the following constraints to $solver$
43:      $(\forall n_1) \cdots (\forall n_m).ReConstr$

44: RefineResult $\leftarrow solver.check()$

On the other hand, the result of QF1.Refine(QF2, 10) is *False*, which indicates the equivalence relation doesn't hold between QF1 and QF2. The counter example actually presents a solution which satisfies the constraints of QF1 but doesn't satisfy the constraints of QF2. "_d_" indicates the data item on the corresponding port indicated by the capitalized letter while "_t_" indicates the time moment. Time-related constraints for well-definedness are clearly satisfied. The data streams in this specific counter example satisfy the data-related constraints in both QF1 and QF2. However, the time stream on the three output ports of QF2 should be exactly equal, which are not satisfied. Time constraints in QF1 are relaxed a bit. Only delay of the data item transfer in three *BufferSer* needs to be satisfied and this counter example provides a feasible and correct solution.

```
 1  True, None
 2  False
 3  A_d_0 = 0, A_d_1 = 0, A_d_2 = 0, A_d_3 = 0, A_d_4 = 0,
 4  A_d_5 = 0, A_d_6 = 0, A_d_7 = 0, A_d_8 = 0, A_d_9 = 0;
 5  A_t_0 = 0, A_t_1 = 3, A_t_2 = 6, A_t_3 = 9, A_t_4 = 12,
 6  A_t_5 = 15, A_t_6 = 18, A_t_7 = 21, A_t_8 = 24, A_t_9 = 27;
 7  B_d_0 = 0, B_d_1 = 0, B_d_2 = 0, B_d_3 = 0, B_d_4 = 0,
 8  B_d_5 = 0, B_d_6 = 0, B_d_7 = 0, B_d_8 = 0, B_d_9 = 0;
 9  B_t_0 = 1, B_t_1 = 4, B_t_2 = 7, B_t_3 = 10, B_t_4 = 13,
10  B_t_5 = 16, B_t_6 = 19, B_t_7 = 22, B_t_8 = 25, B_t_9 = 28;
11  C_d_0 = 0, C_d_1 = 0, C_d_2 = 0, C_d_3 = 0, C_d_4 = 0,
12  C_d_5 = 0, C_d_6 = 0, C_d_7 = 0, C_d_8 = 0, C_d_9 = 0;
13  C_t_0 = 2, C_t_1 = 5, C_t_2 = 8, C_t_3 = 11, C_t_4 = 14,
14  C_t_5 = 17, C_t_6 = 20, C_t_7 = 23, C_t_8 = 26, C_t_9 = 29;
15  D_d_0 = 0, D_d_1 = 0, D_d_2 = 0, D_d_3 = 0, D_d_4 = 0,
16  D_d_5 = 0, D_d_6 = 0, D_d_7 = 0, D_d_8 = 0, D_d_9 = 0;
17  D_t_0 = 2, D_t_1 = 5, D_t_2 = 8, D_t_3 = 11, D_t_4 = 14,
18  D_t_5 = 17, D_t_6 = 20, D_t_7 = 23, D_t_8 = 26, D_t_9 = 29,
```

Example 3. Consider a travel service scenario which involves reserving hotel and booking transportation (flight or train). The service package processes clients' requests in the following way: The service is initiated through a request from some client, then the request is first handled by the hotel service through a *SyncSer*. Next the request is further transferred to the transportation service, where the *RouterSer* operates and sends it to the flight service and train service. Meanwhile, flight and train booking are both monitored by a government service, which is aggregated through a *MergeSer*. The transportation reservation succeeds only if the government service accepts the reservation. Besides, the hotel service and the transportation service are composed through a sequential composition. This service package can be modeled in Z3 as follows.

```
 1  Serv1 = Service('Sync', 'True', ('A', 'B'))
 2  Serv2 = Service('Router', 'True', ('C', 'D', 'E'))
 3  TravelPack1 = Cloud()
 4  TravelPack1.compose('SeqComp', Serv1, Serv2, 1)
 5  TravelPack1.config('Merger', 'True', 'D', 'E', 'F')
```

Another service package can be simpler, which involves a hotel service and a government service in charge of transportation reservation. Moreover, these two services are also composed through a sequential composition. The model of this simplified service package is presented in the following.

```
1  Serv1 = Service('Sync', 'True', ('A', 'B'))
2  Serv3 = Service('Sync', 'True', ('C', 'F'))
3  TravelPack2 = Cloud()
4  TravelPack2.compose('SeqComp', Serv1, Serv3, 1)
```

Intuitively, these two service packages provide the clients with the same result: hotel and transportation reservation. Therefore, next we check if the equivalence relation exists between them. After invoking the **Refine** function in two directions, we get the returned results shown as follows:

```
1  True, None
2  True, None
```

The result shows that the *TravelPack1* is indeed a refinement of *TravelPack2* while the refinement relation holds also in the other direction. Finally, the equivalence relation between them gets proved.

5 Conclusion and Future Work

This paper extends our previous work on the design model for cloud services and proposes a framework on formal specification of cloud services and compositions in SMT solver Z3. The composition of cloud services is given by a family of composition operators which are specified in a class capturing the behavior of the composition. The framework naturally preserves the original choreography of cloud applications, and thus makes the description of cloud services and applications reasonably readable. This work also provides a complement of the verification by theorem proving approach in our previous work. In fact, sometimes users of theorem provers like Coq or PVS need to construct proofs or even build counter examples manually first to show that a property is not satisfiable. For such cases, using Z3 makes it possible to automatically search for possible bounded counter examples.

However, the proposed framework focuses on addressing the data- and time-related properties of the top level cloud services while failing to address the availability and failure rate of online services. The case studies are also set in a high level conceptual setting. In future work, we plan to incorporate the QoS aspects on cloud services into this model and will investigate the formalization and quantitative reasoning about low level programs of cloud applications in SMT solvers. On the other hand, we hope to develop the formal model for dynamic reconfiguration and adaptation of cloud services as well, which is quite useful in real world scenarios.

Acknowledgement. The work was partially supported by the National Natural Science Foundation of China under grant no. 61772038 and 61532019.

References

1. Aceto, L., Larsen, K.G., Morichetta, A., Tiezzi, F.: A cost/reward method for optimal infinite scheduling in mobile cloud computing. In: Braga, C., Ölveczky, P.C. (eds.) FACS 2015. LNCS, vol. 9539, pp. 66–85. Springer, Cham (2016). https://doi.org/10.1007/978-3-319-28934-2_4
2. Alt, L., et al.: HiFrog: SMT-based function summarization for software verification. In: Legay, A., Margaria, T. (eds.) TACAS 2017. LNCS, vol. 10206, pp. 207–213. Springer, Heidelberg (2017). https://doi.org/10.1007/978-3-662-54580-5_12
3. Benzadri, Z., Belala, F., Bouanaka, C.: Towards a formal model for cloud computing. In: Lomuscio, A.R., Nepal, S., Patrizi, F., Benatallah, B., Brandić, I. (eds.) ICSOC 2013. LNCS, vol. 8377, pp. 381–393. Springer, Cham (2014). https://doi.org/10.1007/978-3-319-06859-6_34
4. Chen, L., Fan, G., Liu, Y.: Modeling and analyzing cost-aware fault tolerant strategy for cloud application. In: Proceedings of SEKE 2016, pp. 439–442. KSI Research Inc. and Knowledge Systems Institute Graduate School (2016)
5. de Moura, L., Bjørner, N.: Z3: an efficient SMT solver. In: Ramakrishnan, C.R., Rehof, J. (eds.) TACAS 2008. LNCS, vol. 4963, pp. 337–340. Springer, Heidelberg (2008). https://doi.org/10.1007/978-3-540-78800-3_24
6. Fitch, D., Xu, H.: A raid-based secure and fault-tolerant model for cloud information storage. Int. J. Softw. Eng. Knowl. Eng. 23(05), 627–654 (2013)
7. Freitas, L., Watson, P.: Formalizing workflows partitioning over federated clouds: multi-level security and costs. Int. J. Comput. Math. 91(5), 881–906 (2014)
8. Graiet, M., Mammar, A., Boubaker, S., Gaaloul, W.: Towards correct cloud resource allocation in business processes. IEEE Trans. Serv. Comput. 10(1), 23–36 (2017)
9. Hoare, C.A.R., He, J.: Unifying Theories of Programming. Prentice Hall International, Upper Saddle River (1998)
10. Jifeng, H., Li, X., Liu, Z.: rCOS: a refinement calculus of object systems. Theor. Comput. Sci. 365(1–2), 109–142 (2006)
11. Laibinis, L., Byholm, B., Pereverzeva, I., Troubitsyna, E., Eeik Tan, K., Porres, I.: Integrating Event-B modelling and discrete-event simulation to analyse resilience of data stores in the cloud. In: Albert, E., Sekerinski, E. (eds.) IFM 2014. LNCS, vol. 8739, pp. 103–119. Springer, Cham (2014). https://doi.org/10.1007/978-3-319-10181-1_7
12. Nawaz, M.S., Sun, M.: Using PVS for modeling and verifying cloud services and their composition. In: Proceedings of CBD 2018, pp. 42–47. IEEE (2018)
13. Reddy, G.S., Feng, Y., Liu, Y., Dong, J.S., Sun, J., Kanagasabai, R.: Towards formal modeling and verification of cloud architectures: a case study on Hadoop. In: Proceedings of SERVICES 2013, pp. 306–311. IEEE Computer Society (2013)
14. Sim, K.M.: Agent-based cloud computing. IEEE Trans. Serv. Comput. 5(4), 564–577 (2012)
15. Sun, M., Arbab, F., Aichernig, B.K., Astefanoaei, L., de Boer, F.S., Rutten, J.J.M.M.: Connectors as designs: modeling, refinement and test case generation. Sci. Comput. Program. 77(7–8), 799–822 (2012)
16. Sun, M., Fu, G.: A formal design model of cloud services. In: Proceedings of SEKE 2017, pp. 173–178. KSI Research Inc. and Knowledge Systems Institute (2017)
17. Zhang, P., Lin, C., Ma, X., Ren, F., Li, W.: Monitoring-based task scheduling in large-scale SaaS cloud. In: Sheng, Q.Z., Stroulia, E., Tata, S., Bhiri, S. (eds.) ICSOC 2016. LNCS, vol. 9936, pp. 140–156. Springer, Cham (2016). https://doi.org/10.1007/978-3-319-46295-0_9

On Efficiency of Scrambled Image Forensics Service Using Support Vector Machine

Sahibzada Muhammad Shuja[1], Raja Farhat Makhdoom Khan[1],
Munam Ali Shah[1], Hasan Ali Khattak[1(✉)], Assad Abbass[1],
and Samee U. Khan[2]

[1] COMSATS University Islamabad, Islamabad, Pakistan
hasan.alikhattak@gmail.com
[2] North Dakota State University, Fargo, ND, USA

Abstract. Images can be a very good evidence during investigation of a crime scene. At the same time they can also contain very personal information which should not be exposed without the consent of the involved people. In this paper, We have presented here a practical approach to protect privacy of under investigation images with the use of Arnold's Transform (AT) scrambling and Support Vector Machine, We also provide a new approach towards the whole forensics service provided by the designated agencies with the help of implementation of our approach. We enhanced the security of AT and provided privacy preserving mechanism to ensure protection of privacy. In literature only policies are defined to protect the privacy and lack of a solid approach which we have tried to resolve with a proof of concept implementation. In short, we have provided a full image forensics framework for illegal image detection while preserving the privacy.

Keywords: Digital forensics · Privacy · ORB · Arnold's Transform · SVM · Guns trafficking · Image forensics

1 Introduction

In today's world, computer and other devices like smart phones are in use in a great number which also increase their involvement in the crimes. Hackers and fraudsters are finding these as easy tools for committing the crime. Digital forensics or computer forensics tries to search the evidence from those devices so criminals can be charged in a court in a legal manner. Forensics is a significant service in maintaining law and order situation. In a forensics process, involved device is taken and a full copy of the device hard disk is taken for further analysis. There is no doubt of the good work that is done by the computer forensics but there are many files which a forensics team get that are not relevant for the investigation. Another serious privacy matter arises that when there

Y. Xia and L.-J. Zhang (Eds.): SERVICES 2019, LNCS 11517, pp. 16–30, 2019.
https://doi.org/10.1007/978-3-030-23381-5_2

comes the chance that an investigated person may not be turn up as a convict, compromising his privacy in vane. Any file can reveal the privacy of a person but none more than image files that can be very personal and reveal of them causing a serious privacy breach scenario. Now, people are very aware and conscious about their privacy and even don't want authorities to get involved in it.

Images can be very helpful not only prosecuting someone for the possession of illegal image but can also identify a person involvement in some very illegal businesses like arms trafficking and child pornography. Digital image forensics is a vast field working in identification of source of an image with the help of image features, individual source identification and identification of modifications. We can also say that image forensics involves photogrammetry, photographic comparison, content analysis and image authentication. Our focus is on photographic comparison and content analysis of the image. The scenario that we will be going to follow is of keeping images of guns that can indicate the possession of illegal images, guns trafficking or though a slight chance but can also prove involvement in a murder.

Currently image privacy and security is preserved with the means of two techniques, one is encryption of image into a meaningless form and second one is scrambling which just scramble around the pixels of image so a human eye can not identify the contents of the image. Scrambling is a lot better than encryption in way of extracting features. Arnold's Transform [1] is a widely used technique of scrambling and known for preserving features of image and its periodicity making image recoverable again. Features extraction can be done with methods like Scale Invariant Feature Transform (SIFT) [2] and Oriented Fast Rotated Brief (ORB) [3] which combined with Support Vector Machines (SVM) can provide great results.

We have combined all these different powerful techniques to make our proposed technique for privacy preserving image forensics. Our approach is better than the perceptual hashes [4] used before for providing privacy preserving image forensics because of it's robustness & accuracy along with the preservation of security & privacy as well. We will explain our technique in detail in upcoming sections but one key thing to note is that with the help of this, crimes given above can be investigated in a fast and easy way. Our main contributions in this work which are summarized as follow:

- We have improved the already present Arnold's Transform (AT) algorithm which gives privacy but is not very good in terms of security which can lead to the breach of privacy. We bounded it with a key that can give it much better security and we have also increased it's efficiency by modifying it for our approach.
- We have created a new approach from a combination of AT algorithm, ORB and SVM to provide an efficient and privacy preserving method for the image forensics.

– We have explained our technique working according to the forensics process and also provided some privacy policies that should be implemented along with our technique.
– We have just not given a theoretical view but also proved with the help of results that are generated by our experiments.

We have given our related work in Sect. 2 and then provided detail of Enhanced-AT(EAT) in Sect. 3. Our proposed technique is explained in Sect. 4, forensics process combination with privacy policies in Sect. 4.2 and then results & experiments in Sect. 5. Finally, we have concluded our paper in Sect. 6.

2   Related Work

Digital image forensics is an important part of digital forensics that helps in gathering evidence from the images with the help of different tools and is connected with other domains in computer sciences. We have included a detailed related work because the combination of different fields. We have provided related work in our related fields which also shows the importance of our work.

2.1   Privacy in Digital Forensics

Privacy in the digital forensics is a quite interesting topic and some researchers have worked on. In [5] security and privacy is discussed in the computer forensics context. Privacy violation can be here that forensics investigator has access to even all the unrelated files too when a copy of whole device is made. There are different policies given for the investigators that should be followed, one of which is removal of unnecessary data. Another study shows that only policies are not enough for this purpose so in [6] a study was done on the privacy problems in the computer forensic and gives that the collection of only related data should be done like only emails that are related to the topic not thousands of others too. Already published digital forensics investigation models and procedures are not supporting the data privacy protection. Fourth amendment and many other privacy policies are already supporting the privacy but encryption should also be done to protect data in both collection and analysis phase. Here, AES and HMAC like stream ciphers are recommended because they are fast and efficient.

As above is general discussion about the privacy in digital forensics but now there are already a technique present that is used for privacy preserving image forensics in a practical way given in [4] as an architecture for image recognition and maintaining privacy. Two departments of law enforcement agency are working together to catch the people with illegal content. First department starts the investigation by encrypting the collected images and after that gives to the other department which match the extracted perceptual hash values from collected images with the already present database of hash values.

Table 1. A comparison of SIFT (Accuracy 58.28% time = 0.153), ORB (Accuracy 51.95% time = 0.024) and SURF (Accuracy 50.43% time = 0.049) on differently distorted and rotated images

Time/match rate	Varying intensity	Rotated image	Scaling	Shearing	Noisy image	Distortion
SHIFT	0.13 sec 76.7%	0.16 sec 65.4%	0.25 sec 31.8%	0.133 sec 62.89%	0.115 sec 53.8%	0.132 sec 59.09%
SURF	0.04 sec 72.6%	0.03 sec 50.8%	0.08 sec 3.60%	0.049 sec 59.04%	0.059 sec 39.48%	0.036 sec 44.06%
ORB	0.03 sec 63.6%	0.03 sec 46.7%	0.02 sec 49.5%	0.026 sec 51.88%	0.027 sec 54.48%	0.012 sec 46.04%

2.2   Image Encryption (Scramble) Methods

There is not much study until now in the privacy preserving image matching and forensic but there are many image scrambling techniques that are present. The most used is AT or Arnold Cat map first presented in 1960s, image scrambling technology depends on the information hiding technology which provides the techniques of information that don't need passwords. Arnold transform is widely used because of its features like simplicity and periodicity. Arnold transform was only applicable for squared images when it came out in 1960s as Arnolds cat map but in [1] an algorithm for non-square images is provided. There are also phase scrambling technique given in [7] that scrambles images according to the phase co-relation only and match them without any key needed, key is only one time used for scrambling. In [8] there is a comprehensive survey on different image scrambling techniques which includes Rubiks Cubic algorithm, Arnold Transformation, Sudoku Puzzle scrambling and R-Prime shuffle technique. All these techniques can scramble image and then unscramble it easily. Similarly, in [9] different scrambling techniques are presented including Arnolds Transform, Poker Shuffling Transform and Fibonacci Transform. In all papers AT is found very efficient and providing enough security.

2.3   Feature Extraction and Matching Methods

Extracted features can provide very powerful image matching and retrieval in the domain of image processing. This can also be extended to next level of finding the features from scrambled domain that match our requirements. There are many methods but some of most popular of them are SIFT which was first published in 1999 and patented [2]. It uses differences of Gaussians (DoG) and extract local features, in [10] a privacy preserving SIFT (PPSIFT) concept is given which uses a homomorphic encryption technique and do DoG in Paillier encrypted domain. Speeded-Up Robust Features is another technique given in [11] is also patented and it follows same principles and steps like SIFT but is different like it uses square shaped filters as approximation of Gaussian smoothing for detection. Latest one is Oriented FAST and Rotated BRIEF (ORB) developed by Open CV labs and is totally free to use and is much faster than previous techniques as given in [3]. It is a mixture of two techniques FAST (Features from Accelerated Segmentation Test) used to find key points while BRIEF (Binary robust independent features) for visual descriptors. In a latest study [12] there is a comparison of SIFT, Speed Up Robust Feature (SURF) and ORB performance of image matching in a distorted domain. We have combined all in Table 1 to show a combined performance of all three. When machine learning like SVM are applied on feature extraction methods, results can be very good.

3   Enhanced Arnold Transform

Arnold's transform is a key element in the working of this technique because of its ability to provide privacy along with homomorphic making operations possible on scrambled image. Introduced as Arnold's cat map by Vladimir Arnold in 1960, it works with a square image. First the image is sheared one unit up, then two units to the right, and all that lies outside unit square is shifted back by the unit until it is within the square. This is shown in equation below:

$$T = (x, y) \rightarrow (2x + y, x + y) \, mod \, 1 \tag{1}$$

AT is better than other scrambling techniques because of its simple mathematical operations usage and a controlled image scrambling process. It can be seen in (1), how it works. An image de-scrambling can be done in same way as scrambling, running iterations again. When first introduced, it could only be used with square images which was a weak point for it but then in [4] a method for non-square images was proposed. In non-square approach, same square algorithm is used but image is divided into multiple square and each square is scrambled separately. On the other hand, for the purpose of descrambling all squares in the image are descrambled to get the original image.

In Algorithm 1 we have given our method that is used in our proposed technique for scrambling. This algorithm needs an image, a key converted into an integer value and iterations which is also an integer value that tells algorithm that how much scrambling to image should be done. The key here is given to the AT to control it in a secure manner. It can be seen in (1) that a value 2 is given to x and y, instead of 2 here is user choice represented as a key. In this transform to properly work in descrambling again a full same equation is required, so after some iterations original image can be regenerated. So, without the key it is really hard to descramble the image, though a word of caution that very high values of key can scramble image very much and can hide much of the information. Key can have values like 2, 10 or any integer, if some one tries to descramble image without the key he has to use brute force to try each value and also has to wait for many iterations to check if this is the image or not making it a two layer security. Now come to the iterations value which is also user choice and also controls that how much scrambling should be done because if equation runs again and again it will keep scrambling the image and will produce very scrambled image.

Algorithm 1. EAT Algorithm Scramble (Image, Key, Iterations)

```
 1 Start
 2 N; imageArray; xmap; ymap; results;
 3                          ▷ /*Assign values to variables*/
 4 imageArray = convert image to array;
 5 N = imageArray size;
 6 x = imageArray columns;
 7 y = imageArray rows;
 8                          ▷ /*Applying Arnold Transform*/
 9 imageArray = convert image to array;
10 xmap = (Key*x+y)%N;
11 ymap = (x+y)%N;
12                          ▷ /*Scrambling of the image*/
13 for i in the range of N+1 do
14    if i=iteration then
15       | save image;
16    end
17 end
18 imageArray =imageArray[xmap,ymap];
19 end
```

We called it EAT because in original one there was no control over security and efficiency but now it can be controlled. Privacy is given by transform because images become enough scrambled that a human eye can't recognize what is in the image. Security is provided with two things, first one are iterations that are needed to descramble images and second one is now the key which we have provided as a new security measure. The great thing about this scheme is even from the noisy looking image, features can be extracted which we are going to discuss in upcoming section of our proposed scheme. One thing to be noted that if security is high getting features from images becomes difficult, so there is a balance needed between security and usability. We have explained it with results in the results and experiments section making it more understandable.

As for the non square approach our same algorithm will be used but to scrambled each square of the image has to be scramble separately and than rejoined to a single image again. For descrambing, each square inside the image has to be descrambled and then the combined image of all the squares will show original image. It is a lot more complex and slow than the original AT because of taking parts of an image and applying algorithm on them. Because of this complexity it is also more secure than the non-square technique. For readers more interested in AT we recommend [1] and [13].

4   A New Approach for Privacy Preservation

We have created our approach from a combination of already state of the art techniques in this field. There is not much work has been done on the privacy preserving computer forensics except defining different policies. We have chosen SVM, AT and ORB from machine learning, encryption & scrambling domain and feature extraction respectively. We have already explained in detail our EAT so first we are going to provide a little detail about ORB and Support Vector Machine and then we will give our proposed scheme.

4.1   Preliminaries

Support Vector Machine. SVM is a very popular and used machine learning approach like decision trees & neural networks. SVM is used for both classification and regression like other present techniques but its real power is its kernel system. It works by taking a training set which is divided into different categories, SVM algorithm assigns them into different categories. Basically SVM model is representation of examples as points in space, these points are mapped so that separate categories are divided with a clear gap as wide as possible [14]. It can also perform non-linear classification with the help of kernel trick and represent them in a higher dimensional space. With the use of kernels best learning parameters can be found for given data. In case of linear kernel or no kernel boundary will be separated with a straight boundary and requires more samples for better accuracy. In a kernel approach boundary can also non linear like circular or even more irregular one and requires less samples for producing same accuracy as a linear one. Radial Based Kernel or RBF is commonly used in SVM and have two parameter one is called as gamma and other one C. Gamma can be thought as spread of the kernel, If gamma is low than the decision boundary will be broad but when gamma is high decision boundary is low. On the other hand C is a penalty parameter, if low classifier will be okay with the incorrect classifications and then other way around. We have chosen technique because of its efficient performance and ability to handle different type of data especially image features in the form of visual bag of words.

Oriented FAST and Rotated BRIEF. Oriented FAST and Rotated BRIEF simple called as ORB [3] is a feature extraction and description technique. Like its predecessors SIFT and SURF it performs same functionality but in a fast and accurate manner. It uses FAST for the purpose of detection of keypoints, when keypoints are found Harris corner measure method is applied to find top points among those points. One problem which was there that FAST is not compute the orientation of the points, so authors used intensity weighted centroid computation for the patch that have corner located at center. So, now the direction from the corner point to centroid gives orientation. On the next stage are descriptors used for extracted features description, ORB uses BRIEF descriptors but it was performing poorly with the rotation, so for solving this problem it is steered according to the orientation of keypoints. This is much faster than SIFT [2] because of not using Difference of Gaussian which is very slow and uses all the points which are in most cases are unnecessary. In Fig. 1, we have compared SURF, SIFT & ORB which shows that ORB is very fast as compare to SIFT and is very little less in accuracy than SIFT. Accuracy of ORB is increased a lot with the use of SVM and due to its robustness it will be helpful in analyzing a big number of images in a little time. For readers interested in more detailed working of ORB we recommend [3].

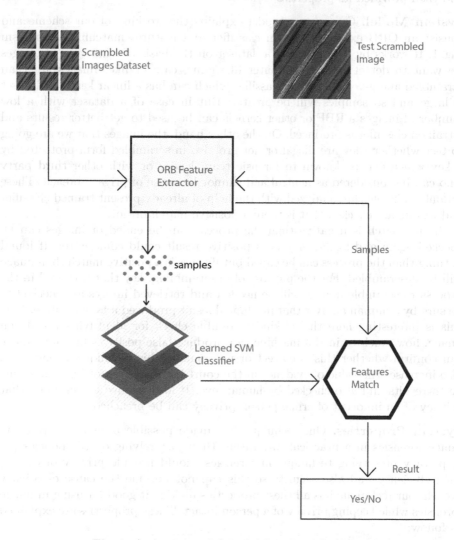

Fig. 1. A system model explaining the technique

4.2   Proposed Scheme

Now, lets come to the scheme that we have proposed for solving the privacy issue in image forensics. We have used EAT for scrambling purposes so privacy and security is ensured. Then we have used SVM to train classifier with the features extracted from the scrambled images and then different scrambled images are given for test that if there is an image present with same features can be detected. We have explained working of our whole system with the help of a system model and then provided its properties as follow:

System Model. Our system model explains the working of our scheme and consist an ORB extractor, a svm classifier and features matching mechanism Fig. 1. It works by first choosing a dataset on the basis of what type of images we want to detect from a computer like guns, cars or like. Images are than scrambled and given to a SVM classifier which can has a linear kernel if dataset is large and so samples will be greater. But in case of a dataset with a low number of images a RBF or other kernels can be used to get better results and a trained classifier is produced. On the other hand, the images that we are going to test whether they are illegal or not are also in scrambled form protected by a key which can be known to forensic team leader or with other third party who can be considered as neutral and cannot be cause of privacy breach. These scrambled images are analyzed with the help of already present trained classifier and results are checked that is there a positive match or not.

If the match is negative than the process can be ended or images can be checked again and again, may be a positive result could come latter. If found nothing than the process can be ended but if there is a positive match than image will be descrambled. For the purpose of descrambling, key that was used in the process to scramble images will be needed and retrieved images are checked to be sure by a human analyst that machine doesn't produced a false positive. Now this is interesting here this is kind of double check for identifying the illegal image, how much trained a machine can produce false positives but human eye can confirm whether this is correct or not making this technique strong. This also increases the value of evidence in the court because it is not just a machine given results and also checked by human eye. It also remove the problem that privacy of an innocent of crime person privacy can be breached.

System Properties. Our technique has made possible providing privacy in image forensics in a practical way rather than just relying on the policies [6]. A privacy preserving technique in forensics should provide privacy of course, then efficiency and also security so this can not become the cause of privacy breach, our technique has all these properties making it good for using in image forensics while keeping privacy of a person intact. These properties are explained as follow:

1. **Privacy:** It is of course foremost property that a privacy preserving system should have and privacy is provided in our technique in two ways. First one is with the help of scrambled images which are can not be identifiable by an human eye. Secondly, we have used SVM which is a machine learning technique and unlike an human analyst, machine analyzing the images can not be considered as privacy breach.

2. **Security:** Another powerful aspect of our technique is making sure provision of security which also increases privacy. We have developed an EAT given in

Sect. 3 that provides a double security. At first it is bind with a key which is if not given images can't be descrambled without brute force. But there is a second security measure in AT that many iterations are required to get back an original image [1]. So, attacker has to use brute force for key checking and also have to check image all iterations that original image has been produced or not, making it a lot more difficult.

3. **Efficiency:** All three techniques that combines into our technique are very fast in scrambling, features description and machine learning. This also gives a boost to our technique making it very fast. SVM and ORB combination increases the accuracy of feature matching. So, in few word it can be said that it is very efficient in both accuracy and robustness.

4. **Double Check:** Our approach also provides a double check mechanism, first machine match images but its results can be false positives. So, instead of just relying on it a human can check the images that came as positive. Use of this ends the problem that someone can be falsely accused and also ensures privacy that only images which are flagged as positive are checked not all collected images (Fig. 2).

Fig. 2. Results showing our approach outputs with labels on left and on right unscrambled form of images

5   Experiments and Results

5.1   Experimental Conditions

For the purpose of checking the performance of the scheme we have applied it on different datasets which contain different number of images. The good property of our technique is that with use of ORB it can produce great results with small amount of data. We have created dataset from pistol as one category and other one as other which contains different images like vehicles including aeroplane, cars and bicycles. We did this because we only need to obtain some pistol results not other images present in the user device because this can breach the privacy. We have taken datasets from just 30 images for other and pistol category to 500 images for each category. ORB needs just few total clear images of the element to be found and it produces rather improved results with a combination of SVM Fig. 5.

In our practical implementation, we have used Python for the purpose of image processing and machine learning. We used libraries like sci-kit learn, OpenCV, matplotlib and numpy for the purpose of machine learning and image processing. System that we used has core i7 2.4 GHZ processing chip, 16 GB RAM and a 4 GB graphical chip.

Fig. 3. Images scrambled to iteration 3 of the Arnold Transform

5.2   Image Scrambling

We have used our EAT for the scrambling purposes, we will explain our most successful dataset which was consist of 102 images total and with a dimension of 200 px. When all 102 images scrambled to 3 iterations, this whole process took only 3.30 sec and provided results like Fig. 3.

It can be seen that with iteration 3 scrambling a lot of prominent feature are there which can be extracted with the help of ORB. But when we scrambled all images to 4 iterations, time taken was 18.54 s and example of this scrambling is shown in Fig. 4. Along with much time taken there are less prominent features in this much scrambled image. Basically this depends on how much image is scrambled, if image is scrambled too much there will be small number of prominent features to be extracted. On the other hand scrambled image at 3 or 4 iterations can produce some good results.

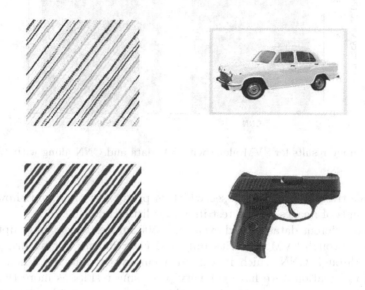

Fig. 4. Images scrambled to iteration 4 of the Arnold Transform

5.3   Final Results

As above some results of scrambling are shown, here we will talk about our final results where SVM and CNN is applied on the scrambled images given for test. With these experiment we concluded that if we will increase the scrambling of images it will be more difficult to obtain good features from the images. By applying CNN, gives unfair results of descrambled image. When SVM is applied all results that come up with label pistol saved in a other folder. We descrambled images through SVM can be easily as compared with CNN come up with label pistol by going to some iterations and got original image back, these results are shown in Fig. 6 where it can be seen clearly that scrambled images are labeled with pistol and other by SVM which saved the results after running the classifier on scrambled images given for testing. On the other side there are original images which shows that in other is a bicycle and car as other category contained aeroplanes, bicycles and cars. Similarly, there is also pistol descrambled image which represent the second training category pistol as an example of gun. This also has given very fast results and test 11 images in

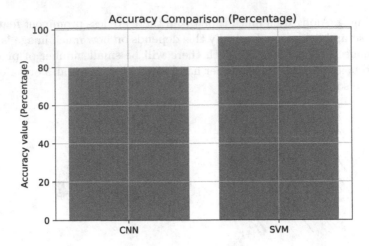

Fig. 5. Accuracy results for SVM along with 500 data and CNN along with 102 data

0.2 sec while training took only 4 sec which is pretty fast time and shows that how thousands of images can be test in a very little time.

We used different datasets and even with 500 images we got only upto 96% of accuracy through SVM but with only 102 clear images got accuracy more than 80% through CNN which is a good accuracy percentage. When images scrambled to iteration 3 we have got very promising results as more than 96% accuracy but in case of iteration 4 scrambled images we could only get up to 70% accuracy point which is a low accuracy point. We have shown our accuracy

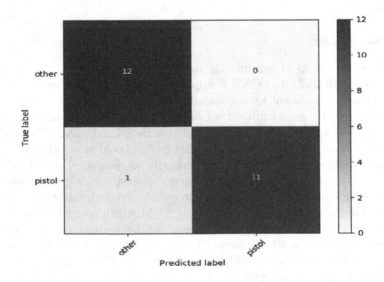

Fig. 6. Confusion Matrix showing accuracy of our approach

in the form of a confusion matrix in Fig. 6. Here the diagonal elements represent the number of points which are predicted, truly while elements off the diagonal represents the elements that are misplaced by the SVM classifier. Figure 5 show the graphical view of comparison between CNN and SVM classifiers in term of accuracy achieved by them.

6 Conclusion

We have proved that it is possible to preserve privacy in a practical manner and with automation becomes more efficient. We have given a full model of our proposed technique that how it works and whole forensics process explanation based on a scenario which shows that this is really applicable in practical use. There are also some policies for our technique are defined which can protect user privacy in a much better form. All in one our technique is a full framework that can be used for the purposes of identification of illegal images with preserving user privacy. This can provide very fast results which can help in fast progression of cases like guns trafficking and child pornography. But it is very difficult to provide privacy in computer forensics investigations unless a persons that are responsible for implementing this technique follow moral values. But it is a lot better than just implementing policies for the investigators because it is a solid approach. We have basically showed that it is possible to provide privacy in computer forensics in a practical way.

Acknowledgment. The work of Samee U. Khan is based upon work supported by (while serving at) the National Science Foundation. Any opinion, findings, and conclusions or recommendations expressed in this material are those of the authors and do not necessarily reflect the views of the National Science Foundation.

References

1. Min, L., Ting, L., Yu-Jie, H.: Arnold transform based image scrambling method. In: Proceedings of 3rd International Conference on Multimedia Technology (ICMT 2013) (2013). https://doi.org/10.2991/icmt-13.2013.160
2. Lowe, D.: Object recognition from local scale-invariant features. In: Proceedings of the Seventh IEEE International Conference on Computer Vision (1999). https://doi.org/10.1109/iccv.1999.790410
3. Rublee, E., Rabaud, V., Konolige, K., Bradski, G.: ORB: an efficient alternative to SIFT or SURF. In: 2011 International Conference on Computer Vision (2011). https://doi.org/10.1109/iccv.2011.6126544
4. Peter, A., Hartmann, T., Muller, S., Katzenbeisser, S.: Privacy preserving architecture for forensic image recognition. In: 2012 IEEE International Workshop on Information Forensics and Security (WIFS) (2012). https://doi.org/10.1109/wifs.2012.6412629
5. Srinivasan, S.: Security and privacy in the computer forensics context. In: 2006 International Conference on Communication Technology (2006). https://doi.org/10.1109/icct.2006.341936

6. Law, F.Y., et al.: Protecting digital data privacy in computer forensic examination. In: 2011 Sixth IEEE International Workshop on Systematic Approaches to Digital Forensic Engineering (2011). https://doi.org/10.1109/sadfe.2011.15
7. Ito, I., Kiya, H.: Phase scrambling for blind image matching. In: 2009 IEEE International Conference on Acoustics, Speech and Signal Processing (2009). https://doi.org/10.1109/icassp.2009.4959885
8. A comprehensive survey on image scrambling techniques. Int. J. Sci. Res. (IJSR) **4**(12), 813–818 (2015). https://doi.org/10.21275/v4i12.nov152034
9. Shelke, R., Metkar, S.: Image scrambling methods for digital image encryption. In: 2016 International Conference on Signal and Information Processing (IConSIP) (2016). https://doi.org/10.1109/iconsip.2016.7857449
10. Hsu, C., Lu, C., Pei, S.: Image feature extraction in encrypted domain with privacy-preserving SIFT. IEEE Trans. Image Process. **21**(11), 4593–4607 (2012). https://doi.org/10.1109/tip.2012.2204272
11. Bay, H., Tuytelaars, T., Van Gool, L.: SURF: speeded up robust features. In: Leonardis, A., Bischof, H., Pinz, A. (eds.) ECCV 2006. LNCS, vol. 3951, pp. 404–417. Springer, Heidelberg (2006). https://doi.org/10.1007/11744023_32
12. Karami, E., Prasad, S., Shehata, M.: Image matching using SIFT, SURF, BRIEF, and ORB: performance comparison for distorted images. In: Proceedings of the 2015 Newfoundland Electrical and Computer Engineering Conference. St. Johns, Canada (2015)
13. Wu, L., Zhang, J., Deng, W., He, D.: Arnold transformation algorithm and anti-arnold transformation algorithm. In: 2009 First International Conference on Information Science and Engineering (2009). https://doi.org/10.1109/icise.2009.347
14. Support vector machine, 21 June 2018. https://en.wikipedia.org/wiki/Support_vector_machine
15. Baryamureeba, V., Tushabe, F.: The enhanced digital investigation process model. In: The Digital Forensic Research Conference (2004)

Investigating Personally Identifiable Information Posted on Twitter Before and After Disasters

Pezhman Sheinidashtegol[✉], Aibek Musaev, and Travis Atkison

The University of Alabama, Tuscaloosa, AL 35487, USA
psheinidashtegol@crimson.ua.edu

Abstract. Through social media, users may consciously reveal information about their personality; however, they could also unintentionally disclose private information related to the locations of where they live/work, car license plates, signatures or even their identification documents. Any disclosed information can endanger an individual's privacy, possibly resulting in burglaries or identity theft. To the best of our knowledge, this paper is the first to claim and demonstrate that people may reveal information when they are vulnerable and actively seek help; evidently, in the event of a natural disaster, people change their behavior and become more inclined to share their personal information. To examine this phenomenon, we investigate two hurricane events (Harvey and Maria) and one earthquake (Mexico City) using datasets obtained from Twitter. Our findings show a significant change in people's behavioral pattern in the disaster areas, regarding tweeting images that contain Personally Identifiable Information (PII), before and after a disaster event.

Keywords: Social media · Privacy · Disasters ·
Information disclosure · Hurricane · Earthquake

1 Introduction

Social media has succeeded in creating a very convenient platform for individuals to connect with others regardless of their geographic location. As the number of social network users grows each day, the concern for privacy also increases [1–3]. Social media is a significant part of modern life; therefore, it is critically important to address potential privacy concerns, and more specifically the risks that accompany the direct or indirect disclosure of personally identifiable information to the public.

1.1 Personally Identifiable Information (PII)

According to the National Institute of Standards and Technology (NIST) [4], PII is any piece of information that could be used to distinguish or trace a person's

© Springer Nature Switzerland AG 2019
Y. Xia and L.-J. Zhang (Eds.): SERVICES 2019, LNCS 11517, pp. 31–45, 2019.
https://doi.org/10.1007/978-3-030-23381-5_3

identity, e.g., full name, driver's license number, or street address. Additionally, PII is any information that is linked or linkable to an individual, such as license plate, signature, or handwriting [4].

When social users reveal PII, their information can be used against them or their family and friends. The worst-case scenario reveals itself when a professional malicious user acquires an individual's private information. In 2010, a *MythBusters* show host accidentally revealed his PII when he posted a geotagged image of his truck and house on Twitter. The posted geotagged photo of his Land Cruiser in the driveway when accompanied with the text, "Now it's off to work," provided all of the information needed for potential thieves to know the location of the house and that it was at the time uninhabited [5]. In the same year, New Hampshire Police investigation of a series of eighteen burglaries revealed a strong connection between the homeowners' posts on social media and subsequent burglaries [6]. According to the Department of Justice, Bureau of Justice Statistics (BJS), an estimated 16.6 million people, or 7% of all persons 16 or older in the U.S., experienced at least one incident of identity theft totaling 24.7 billion in 2012 [7]. The losses of PII, sensitive and non-sensitive, in the U.S. are prevalent, with serious consequences to individuals and organizations.

Common privacy incidents following serious emergency situations include the phenomenon of targeted spearphishing attacks and online scams asking for money to support "relief efforts" [8]. Another less researched phenomenon is the frequency of people who elect to share PII with anonymous online users after a disaster in an effort to receive short term relief.

Our goal is to investigate if PII related images on Twitter occur more frequently post disasters, such as in the events surrounding earthquakes, hurricanes or any other life-threatening events, where people tend to think less about their privacy. In other words, we expect people to deliberately post more private information during disasters on social media by uploading images of their house or neighborhood to show the damages to their property.

As a matter of fact, the majority of pictures taken with digital cameras or smartphones include longitude and latitude coordinates in the image files stored on those devices [9]. Fortunately, social media platforms, such as Twitter and Facebook, have eliminated most of the georeferenced metadata found in image files [10]. However, within a photo's raster layer, some information related to the user's location can be revealed, i.e. an easily recognizable building, distinguishable neighborhood, street or business sign. Moreover, tweeted images may contain more sensitive information when they illustrate any form of identification, such as a driver's card, a university student/faculty ID or an office badge.

We chose Twitter as the social media platform. Therefore, we should acknowledge that, in general, 69% of adults in the US are active on one or more social media platforms. Yet, only 24% of this population are on Twitter [11]. Furthermore, Twitter users tend to be younger and have higher levels of education than the actual population [12]. Despite all of these facts, Twitter offers invaluable social data which is accessible using a Twitter API.

The rest of the paper is organized as follows. Section 2 discusses related works. Section 3 presents an overview of the dataset collection, limitations of the proposed research, and methodology used for sampling and analysis. Section 4 illustrates results using graphs and figures specific for each of the three majors disasters; and Sect. 5 discusses the correlations between the results and real-world reports of social media and emergency situation incidents. Finally, Sect. 6 poses the risks and rewards of using social media during disasters and explores the potential future work.

2 Related Work

Though there are many discussions to have regarding complications of using social networking services for communication, we distinguish between (i) issues focusing on social media security and privacy limitations, (ii) the influence of social media during disasters, and (iii) the drawbacks of excessive use of social media after disasters.

2.1 Privacy and Security Issues in Social Media

In the United States, the Privacy Act of 1974 regulates the collection of personal information by government agencies. Regrettably, there is no overarching federal law regulating private entities [13]. Most social network platforms aim to preserve their clients' privacy as much as possible [14], especially Twitter. Twitter does not require users to provide their real names; instead, it encourages users to create unique pseudonyms with no relation to their real names. When a photo is deleted on Facebook or Instagram, the image and the information carried in the image (URL and shared link) can be accessible for several days due to "photo-deletion delay." Conversely, this is not the case for Twitter which deletes that sensitive information immediately [14].

Regrettably, even with the compliance of anonymity by the users and strict standards set forth by Twitter, a study by Peddenti et al. [15] in 2017 revealed that only $\simeq 6\%$ of Twitter users are truly anonymous. Using multiple social media networks, it is possible to infer 39.9% more personal information via deanonymization and aggregation [2,16,17]. When algorithms and deanonymization fail, information can also be directly revealed by advertisement companies and social media platforms. Recently in 2013, it was reported that the Facebook bug leaked the private contact information of 6 million users [17].

Revealing personal details about your life on social media where everyone can access the information is risky. While using social media, Cai et al. [18] recommended that individuals disguise their attributes (i.e. using encryption) and remove friendship links in order to achieve the "privacy-utility trade-off." Data mining and social media shortcomings are a cause of concern for privacy on the Internet, and sanitizing network data prior to release is necessary [18]. However, this is not easy to implement on regular days, and it gets more complicated during a disaster.

To the best of our knowledge, all researchers investigating the possible security and privacy issues were looking at users' profile attributes and posted texts, and few have used images for social media and privacy correlations.

2.2 Social Media's Influence During Disasters

In a disaster, whether it is natural (earthquake or hurricane), technological (oil spill) or human (terrorism), a lack of communication fuels a crisis [19]. It is commonplace for organizations to willingly use a comprehensive framework for "disaster social media" in order to successfully employ meaningful social media communication in a disaster [19]. Some government, non-profit and news organizations (Salvation Army, The National Weather Service, US Federal Emergency Management Agency) include frameworks, such as the Crisis and Emergency Risk Communication (CERC) model and the Disaster Communication Intervention Framework (DCIF) [19]. Social media emergency frameworks are employed to better communicate with disaster victims and provide feedback in the event of an emergency; but the way social media is utilized in a disaster should be adjusted to reflect changing circumstances. Social media platforms -including Twitter, YouTube, and Facebook- are among the most important two-way mediated channels of communication by officials for interaction with the public before, during and after natural disasters [20]. This can be demonstrated by the social media awareness during the terror attacks on Brussels [20] and again during the 2011 Virginia earthquake, where tweets of information about the earthquake moved across the state faster than seismic waves did [21]. Similarly, social media was invaluable during the Great East Japan Earthquake in 2011 when Tsukuba had major power outages, towers went down and web-enabled phones and smartphones became the primary devices for media access. In that situation, the number of tweets per day of 39 local governments increased tenfold. Consequently, after an exhaustive list of positive implications to use social media during an emergency -especially for preparedness, response, and recovery- social media has notable drawbacks [22]. Examples of skepticism for using social media during a disaster include unconfined or unreliable information, possible technical problems, and the notable collection of privacy concerns [20]. Our critical approach to social media and privacy concerns also takes advantage of the fact that in the times of disasters, affected people post images to reach the rescue teams, officials, and organizations.

2.3 Privacy Issues in Social Media During Disasters

Social media is known for its share of active attacks including stalking, cyberbullying, malvertising, phishing, social spamming, and scamming; yet, post-disaster the passive attacks could arguably be of greater concern. During a disaster, people often take photos to document the cascading events and subsequent sharing of information in such situations can be informative, newsworthy, and therapeutic [23]. However, privacy issues arise on Twitter post-disaster when sensitive information is revealed through images or texts regarding a person's location or personal information, such as cell phone number.

It is not impossible to infer a anonymous Twitter users whereabouts. According to Hecht et al. [24], 66% of Twitter users have a geographic location in the location field of their users' profile, while the rest of users leave the field blank or filled in with a non-geographic location. Even if the location field is not occupied, it is possible to predict the location based on tweet content. Furthermore, despite the fact that less than 1% of all tweets are geo-tagged, algorithms are available to accurately predict the location of a tweet at the city level from a combination of information including: the tweet contents (e.g., place names, hashtags), tweeting behavior-based time zone location (volume of tweets per time unit), and trained location dataset based on the tweet contents (e.g., the dictionary containing dynamically weighted ensemble of locations) [25].

Unlike the disaster itself, privacy issues which accompany them can be avoided. As people voluntarily capture, gather and aggregate information through social media, the result is a very large-scale collection of personal information [23]; and upon deliberation of privacy consequences, every precaution must be taken when sharing images post-disaster. A pattern exists that people may write a quick message immediately after a life-threatening event but it takes a long time for images and videos to be uploaded from cameras to large-scale social forums (such as Twitter and YouTube). Hence, before any information is compromised on social media, users need to remember that they may be able to partially mask their location by carefully avoiding the mentions of geographic places in their posts [25].

When considering the security risks in the tweets' actual text, one's attention should also be on the information residing in the posted photos. While there is a possibility of predicting the users' location in an image, there is also a possibility of retrieving personally identifiable information in the photo.

3   Experimental Design

3.1   Definition of PII in This Study

Based on our interpretation of the NIST-defined PII, described in Sect. 1.1, we define three types of PII images. These are "location disclosure", "personal information disclosure", and "linkable information."

Location Disclosure. An image is labeled as a location-disclosure PII when it contains a complete exterior view of at least one recognizable building. Most of the images which are tagged by location-disclosure include more than one structure in the exterior view. Therefore, the location of these images can be determined by any individual that is familiar with the area. For instance, Fig. 1a shows a flooded apartment complex in Puerto Rico. The street-level location of this picture can be discovered easily by the people that are familiar with that neighborhood.

Personal Information Disclosure. This type of PII images includes all pictures of government/non-government issued identification cards. In addition, any documents that contain at least the full name of an individual are defined as "Personal information disclosure." Fig. 1b shows a tweeted image of a driver's license found after an earthquake in Mexico City.

Linkable Information. The third type of PII images includes any photos containing information that is linked to or can be linkable to an individual. Examples of this category include tweeted images of a ticket to a concert, a signature on a personal bank check, or as shown in Fig. 1c, the photo of a parked car with a visible license plate.

(a) Location disclosure: the address could be recognized by people living in or familiar with the neighborhood

(b) Personal information disclosure: a driver's license (government issued ID)

(c) Linkable information: a parked car with a visible license plate with the tweet text: "My baby is clean now"

Fig. 1. Examples of images posted on Twitter which are considered PII in this study. The images were blurred to protect the users' privacy.

3.2 Data Collection

Twitter data can be downloaded in various ways. Below, two methods are described for the dataset collection of (1) a Hurricane Harvey Twitter dataset and (2) a Hurricane Maria and a Mexico City earthquake Twitter dataset.

Hurricane Harvey. We used the PowerTrack API to amass a comprehensive dataset of geotagged tweets [26]. Using available operators, all tweets within the vicinity of Houston, TX were retrieved within a radius of 24.5 miles from the coordinates of 29.750641 latitude and −95.365851 longitude. These tweets were collected between the dates of July 21th and October 1st, including more than two weeks before and after the hurricane occurred (Fig. 2a).

Table 1. Overview of datasets

	Before disaster	After disaster
Total tweeted images from Houston	1,568	1,573
Total tweeted images from Puerto Rico	12,644	13,191
Total tweeted images from Mexico City	10,782	15,296

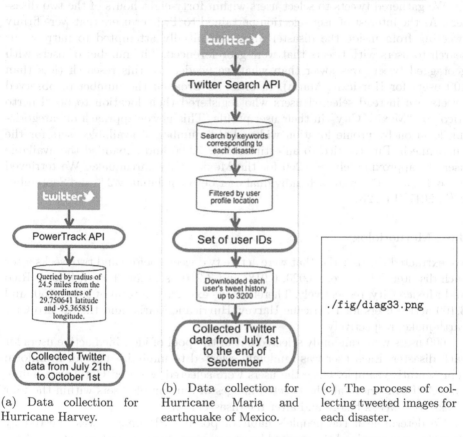

(a) Data collection for Hurricane Harvey.

(b) Data collection for Hurricane Maria and earthquake of Mexico.

(c) The process of collecting tweeted images for each disaster.

Fig. 2. The processes of collecting tweeted images in this paper.

Hurricane Maria and Mexico's Earthquake. Unlike the Hurricane Harvey method, we used the Twitter Streaming API to collect data for Hurricane Maria and the Mexico City earthquake. On Tuesday September 19th at 7:30pm CST, ~11 h prior to Hurricane Maria making landfall on Puerto Rico, tweets were recorded based on designated queries. This recording continued for 48 h. Similarly, tweets posted ~1 h after the Mexico City earthquake (occurring ~6:15pm CST) were recorded based on specific queries. For the Hurricane Maria dataset, we captured all the tweets that mentioned "Hurricane_Maria", "Hurricane", "Maria", "huracan", "Puerto", "Rico" during the Hurricane. For the earthquake dataset, we captured all the tweets that mentioned "earthquake", "Puebla", "Mexico", "terremoto", "sacudida", "shaking", and "Mexico City" at the time of the disaster. These terms were carefully chosen to filter out irrelevant tweets that were not about these particular disasters. In both cases, we used the Spanish equivalent of the English terms, such as "sacudida" and "shaking" for the earthquake in Mexico.

We gathered tweets to select users within forty-eight hours of the two disasters. As the interest of user selection pertained to Twitter users that were firmly tweeting from inside the disaster area, we initially attempted to narrow our search to users with tweets that were georeferenced. The number of users with geotagged tweets was lower than what we needed for this research (less than 100 users for Hurricane Maria). In order to broaden the number of observed tweets, we instead selected users who registered their location to be "Puerto Rico" or "Mexico City" in their user profile. This newer approach of categorizing location by profile location widened the number of available users for the hurricane in Puerto Rico to an estimated $\simeq$ 7,000 and expanded the available users to approximately $\simeq$ 3,200 for the Mexico City earthquake. We retrieved up to 3,200 tweets from each individual user between January 2nd and September 29th, 2017 (Fig. 2b).

3.3   Methodology

We extracted the user IDs that were active two weeks before and two weeks after each disaster. There were 3,920, 6,918, and 3,210 users for Harvey, Puerto Rico and Mexico City, respectively. Therefore, we extracted a total of 985, 6,250, and 3,109 active users for Hurricane Harvey, Hurricane Maria, and the Mexico City earthquake, respectively.

600 users were randomly selected out of the pool of identified active users for each disaster. Each user was randomly selected to minimize bias for a uniform representative sample. Once the users were selected, we used Get Tweet Timelines API to retrieve all the tweeted images of our sample users within the four weeks (two weeks before and after the disaster) [27].

To determine if the people's habit of posting PII images was affected by disasters, we examined the tweeted images of every user for each disaster: before, during, and after it. Table 1 shows the number of tweeted images for each disaster from the streamlined selection of users. Finally, we manually analyzed each of these images to label them as either *PII* or *non-PII* images (Fig. 2c).

4   Experimental Results

4.1   Hurricane in Houston

Tropical storm Harvey intensified to a category 4 hurricane before making landfall along the middle Texas coastline late August 26, 2017. Houston's metropolitan area faced severe rain and wind between the 25th and 29th of August. However, Tweets peaked on August 27th, but this surge in social media presence did not last. After three days, the number of tweets returned to normal where the pattern of daily tweets two weeks after the hurricane mimicked the patterns before the hurricane took place (Fig. 3).

The daily estimate of tweeted images fluctuated but trends emerged for Twitter user presence and behavior throughout the disaster period. While the amount of Twitter users accessing their Twitter accounts and uploading PII images increased by 514% (Fig. 4), the amount of tweeted PII images followed a similar trend, increasing by 633% (Fig. 4) during/after Hurricane Harvey made landfall. There were only four tweeted images (out of 1,568) before the hurricane made landfall, but after the hurricane reached Texas's coastline, this number increased to twenty-three (out of 1,573). Moreover, from a sample of 600 Twitter users, only three Twitter users uploaded PII images to social media compared to the twenty-three users who tweeted PII images during and after the disaster. There was no overlap in users before and after the hurricane.

To examine the change in the pattern of the amount of tweets containing PII images, we performed a paired t-test on the number of PII images posted by each user in our sample group before and after the hurricane. The averaged PII tweeted images during/after the hurricane was more than the average PII images tweeted before the hurricane made landfall (df $= 598$, p $= 0.004$). This suggests that as user presence increased, so did the amount of PII imaged tweets posted per individual.

4.2   Hurricane in Puerto Rico

Puerto Rico was devastated by Hurricane Irma, category 3, on September 6th and again by the rain shield of Hurricane Maria, category 5, on September 20th, 2017. These disasters had significant effects on social media. The two peaks in Fig. 5 indicate the influence that the devastation of the two different hurricanes had on Twitter user behavior. Comparing two time periods - before the hurricanes and the fourteen days during the hurricanes (September 5th to September 19th) before blackouts occurred - on average, the number of tweeted images per day increased by 20%.

To assess the behavioral change of Twitter users posting PII images in a time proximity to the hurricanes, we compiled each posted image containing PII for each user in our sample group before Hurricane Irma and after Irma up until during Hurricane Maria (over 25k images). There was a 388% increase in the number of users posting PII images after/during the disasters (Fig. 6). Subsequently, in the same period of time, the number of posted PII images increased by 276%. To examine the behavioral pattern of users tweeting images containing PII before and after Hurricane Irma, we performed a paired t-test on the number of PII images posted by each user in our sample group before and after the disaster. There were significantly more tweeted images containing PII after the disasters in Puerto Rico (df $= 598$, p $= 5.788e-08$). Similar to Hurricane Harvey, in the period surrounding Hurricane Irma and Hurricane Maria, the user presence increased, and so did the amount of tweets each user posted.

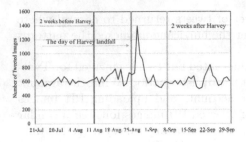

Fig. 3. Daily tweeted images in the Houston metropolitan area. Lines indicate Twitter collection time frame for two weeks before Hurricane Harvey made landfall, when Hurricane Harvey made landfall, and then two weeks after the hurricane made landfall.

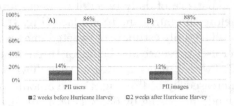

Fig. 4. Tweeted images containing PII over a 4 week period in Houston: (A) Comparison of the number of PII users, (B) Comparison of the number of PII tweets.

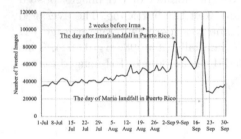

Fig. 5. Daily amount of tweeted images for the island of Puerto Rico two weeks before Hurricane Irma made landfall, the day after Hurricane Irma made landfall, and the day that Hurricane Maria made landfall.

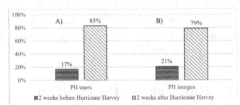

Fig. 6. Tweeted images containing PII over a 4 week period in Puerto Rico: (A) Comparison of the number of PII users, (B) Comparison of the number of PII tweets.

4.3   Earthquake in Mexico City

On September 19th, a 7.1 magnitude earthquake struck central Mexico, in proximity to its capital, Mexico City. In its wake, the earthquake had subtle effects on Twitter user presence and behavior. After examining over 25,000 posted images within a 4 week time-frame, the amount of posted images increased rapidly post-disaster. The number of tweeted images escalated on September 19 (the day of the earthquake) and peeked on September 20, with those two days comprising 27% of the tweets posted for the two weeks following the first seismic event (Fig. 7). For the first two weeks following the disaster, the user presence increased by 355% and the amount of tweeted images containing PII increased by the same percentage; thus, the amount of users directly corresponded to the number of tweeted images containing PII (Fig. 8).

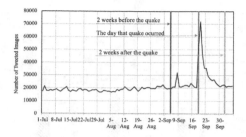
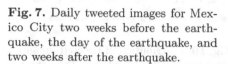

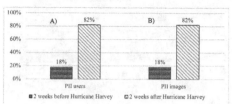

Fig. 7. Daily tweeted images for Mexico City two weeks before the earthquake, the day of the earthquake, and two weeks after the earthquake.

Fig. 8. Tweeted images containing PII over a 4 week period in Mexico City: (A) Comparison of the number of PII users, (B) Comparison of the number of PII tweets.

Using a paired t-test, it was demonstrated that earthquake disasters changed the way affected users managed their PII online. For the Mexico earthquake dataset, we saw that from a 600 user sample, the amount of tweeted PII images after the earthquake was higher than the amount of tweeted PII images before the earthquake (df $= 598$, p $= 6.679e{-}05$).

4.4 PII Image Predominance in Each Disaster

Using three defined categories of PII in Sect. 1.1, i.e. location disclosure, personal information, and linkable information, we distinguished which types of sensitive information these users were revealing through images for each corresponding disaster. Starting with the most prevalent form of sensitive information disclosure, "location disclosure." "Location disclosure" increased after Hurricane Harvey, Hurricane Maria, and the earthquake of Mexico City by 2100%, 457%, 4600%, respectively (Fig. 10). Images with "location disclosure" occurred at the highest frequency compared to the two other types of PII images, though "personal disclosure" was also high after both hurricanes and the earthquake in Mexico City.

Only in two cases of tweeted images from Hurricane Harvey and Maria were there no forms of "personal information" disclosure. Yet, "personal information" disclosure was not as relevant for the earthquake disaster dataset. In fact, after the earthquake in Mexico City, it was only 50% more likely to see personal information being disclosed in a tweeted image (Fig. 10). "Linkable information" did increase after the disasters in Puerto Rico and Mexico City by 200% and 400%, respectively, but while "linkable information" was prevalent for these two disasters, the chance of having tweeted images containing linkable information decreased 33% in the case of Hurricane Harvey (Fig. 10).

For Hurricane Harvey, the authors computed the increase ratio of posted images and the increase ratio of gauge height and mapped them, shown in Fig. 9. The ratio for both gauges and number of images was calculated by $((X_{Afterdisaster} - X_{Beforedisaster})/X_{Beforedisaster})$. The data for gauge height in

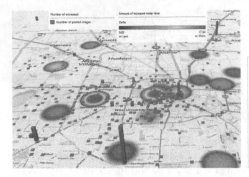

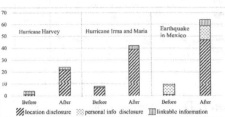

Fig. 9. Distribution of tweeted images based on geographical locations and water intensity.

Fig. 10. Distribution of tweeted images containing PII based on the three defined types of PII disclosure mentioned in Sect. 1.1 - for Hurricane Harvey, Hurricanes Irma and Maria, and the earthquake in Mexico City.

Houston was acquired from the USGS website [28], and the number of tweets computed for each 0.5-minute grid cell is presented as a bar at the center of the cells.

5   Discussion

Users tend to reveal more PII during and after the time of disasters. In fact, people experiencing unexpected natural disasters tend to post more images during and after disasters. Each disaster demonstrated that the number of posted images returned to normal in less than a week. However, in the case of Hurricane Maria, 100% of Puerto Ricans were left without power [29]; therefore, as shown in Fig. 5, there was a quick and drastic decline in the number of tweeted images right after the hurricane.

Figures 4, 6 and 8 show that along with the rise in the amount of tweeted images with PII, the number of users posting PII images is also increasing. People that are devastated by disasters post images more often in an attempt to get help from their local government or rescue teams [23].

After images posted in social media were assessed based on the three PII categories defined in Sect. 1.1, as expected, during disasters people post significantly more images that may potentially reveal their locations. Surprisingly, in the case of the earthquake in Mexico, results show that most of the images taken before the earthquake contained a large quantity of "personal information disclosure." It may have been linked to the several missing person reports that were posted before and after the earthquake. However, almost no images containing "personal information disclosure" had been found after the hurricanes in Puerto Rico and Houston. In addition, images with "linkable information" were recognized in all three disasters, but this category had prevalence during and after Hurricane Maria and the earthquake in Mexico.

Social media provides important data for first responders and rescue teams [20–22]. However, if such data is assimilated by a malicious user, it can threaten users' privacy or safety. While a post containing an image of a flooded neighborhood could guide rescue teams to the area, it could also be troublesome in the future by revealing a home/work location.

According to Frailing and Harper, [30–33] looting is happening in the wake of natural catastrophes. For example, the rate of burglary increased 200% in the aftermath of Hurricane Katrina in New Orleans. Considering the fact that people between 18–29 are the most active users on social media [11] and also responsible for the largest number of crimes in the United States [7], the likelihood of malicious users taking advantage of innocent posts on social media to select their next target increases.

During disaster situations and emergencies, people tend to be distracted and can more easily fall victim to privacy incidents. Users should delete their posts containing PII after disasters. Alternatively, social media platforms can treat the posts that contain PII the same way as they treat the posts containing graphic violence and adult content. In other words, the site can require users to remove the posts containing PII or at least draw their attention to such posts [34].

Several limitations must be acknowledged. Our datasets were limited to the data which were only available from Twitter. Another drawback was that we were unable to explore the differences in Twitter usage by demographic characteristics and the degree to which the users were in proximity to the disaster. Twitter users usually do not reveal such information (birth-date/age, gender, location's coordinate). Furthermore, access to the full dataset of geo-referenced tweets was only available for Hurricane Harvey. Therefore, for the hurricanes in Puerto Rico and the earthquake in Mexico City, we relied on the datasets we recorded using Streaming API.

6 Conclusion and Future Work

Social media is beneficial to the public in the event of a disaster because it improves community awareness, and government agencies and advocacy groups rely on social media sites (such as Twitter) to communicate information to and with the public. Still, as infrastructure fails and the capacity of law enforcement diminishes, crime rates increase and looting/identity theft can be attributed to information being revealed on social media. In this study, we assessed Twitter usage by the public following three disasters and found that users were trusting social media more often during/after a disaster in three separate scenarios. Randomly selected groups of 600 users revealed more significant amounts of personally identifiable information (PII) during and after disasters. In all "location disclosure" cases, users tended to more frequently indicate street addresses, neighboring buildings, or distinctive landmarks in their uploaded images. Users facing the after-effects of Hurricane Maria and the earthquake in Mexico revealed "linkable information" in the form of signatures on bank checks and images of license plates. In the case of the earthquake in Mexico, people tended to disclose "personal information" by revealing birth dates, job positions, or height of missing loved ones.

The level of PII exposed differed from place to place and varied based on the severity of an event. The hurricanes did not have the same rapid initial Twitter response as the earthquake; this is in part due to the longevity of a hurricane event and the lack of accessibility to power during the event. It is possible that the effects of revealing personal information could have long-lasting effect on privacy that would only emerge months or even years after disasters. Users may be advised to follow up to erase sensitive uploads often at ordinary times, but it is especially important to follow up post-disasters.

Future work includes automation of the process of detecting PIIs in images and development of an application to evaluate the posted images on the users' timeline in real time. Eventually, the suggested system should detect the images containing PII and inform the users about the risks they face.

References

1. EMarketer: number of social media users worldwide from 2010 to 2021 (in billions). Technical report (2017)
2. Siddula, M., Li, L., Li, Y.: An empirical study on the privacy preservation of online social networks. IEEE Access **6**, 19912–19922 (2018)
3. Yin, D., Shen, Y., Liu, C.: Attribute couplet attacks and privacy preservation in social networks. IEEE Access **5**, 25295–25305 (2017)
4. McCallister, E., Grance, T., Scarfone, K.A.: Guide to protecting the confidentiality of Personally Identifiable Information (PII) (2010)
5. Murphy, K.: Web Photos That Reveal Secrets. Like Where You Live, NYTimes (2010)
6. Brunty, J., Helenek, K.: Chapter 3 - investigative uses of social media. In: Brunty, J., Helenek, K. (eds.) Social Media Investigation for Law Enforcement, pp. 41–70. Anderson Publishing, Ltd. (2013)
7. US Bureau of Justice Statistics: Prevalence rate of violent crime in the United States from 2005 to 2016, by age. https://www.statista.com/statistics/424137/prevalence-rate-of-violent-crime-in-the-us-by-age/. Accessed 01 July 2018
8. Kryvasheyeu, Y., et al.: Rapid assessment of disaster damage using social media activity. Sci. Adv. **2**(3), e1500779–e1500779 (2016)
9. Boutell, M., Luo, J.: Beyond pixels: exploiting camera metadata for photo classification. Pattern Recogn. **38**(6), 935–946 (2005)
10. Faiz bin Jeffry, M.A., Mammi, H.K.: A study on image security in social media using digital watermarking with metadata. In: 2017 IEEE Conference on Application, Information and Network Security (AINS), pp. 118–123 (2017)
11. Pew Research Center: Who uses each social media platform. http://www.pewinternet.org/fact-sheet/social-media/. Accessed 01 July 2018
12. Mellon, J., Prosser, C.: Twitter and Facebook are not representative of the general population: political attitudes and demographics of British social media users. Res. Polit. **4**(3), 205316801772000 (2017)
13. Narayanan, A., Shmatikov, V.: Myths and fallacies of "personally identifiable information". Commun. ACM **53**(6), 24 (2010)

14. Liang, K., Liu, J.K., Lu, R., Wong, D.S.: Privacy concerns for photo sharing in online social networks. IEEE Internet Comput. **19**(2), 58–63 (2015)
15. Peddinti, S.T., Ross, K.W., Cappos, J.: User anonymity on Twitter. IEEE Secur. Privacy **15**(3), 84–87 (2017)
16. Wang, P., He, W., Zhao, J.: A tale of three social networks: user activity comparisons across Facebook, Twitter, and foursquare. IEEE Internet Comput. **18**(2), 10–15 (2014)
17. Du, S., et al.: Modeling privacy leakage risks in large-scale social networks. IEEE Access **6**, 17653–17665 (2018)
18. Cai, Z., He, Z., Guan, X., Li, Y.: Collective data-sanitization for preventing sensitive information inference attacks in social networks. IEEE Trans. Dependable Secure Comput. **5971**(c), 1 (2016)
19. Houston, J.B., et al.: Social media and disasters: a functional framework for social media use in disaster planning, response, and research. Disasters **39**(1), 1–22 (2015)
20. Knuth, D., Szymczak, H., Kuecuekbalaban, P., Schmidt, S.: Social media in emergencies how useful can they be. In: Proceedings of the 2016 3rd International Conference on Information and Communication Technologies for Disaster Management, ICT-DM 2016 (2017)
21. Honan: Watch the Virginia earthquake spread across Twitter (2014)
22. Palen, L., et al.: A vision for technology-mediated support for public participation & assistance in mass emergencies & disasters. In: Proceedings of the 2010 ACMBCS Visions of Computer Science Conference, pp. 1–12 (2010)
23. Liu, S.B., Palen, L., Sutton, J., Hughes, A.L., Vieweg, S.: In search of the bigger picture: the emergent role of on-line photo sharing in times of disaster. In: Proceedings of the 5th International ISCRAM Conference, vol. 8, no. May, pp. 140–149 (2008)
24. Hecht, B., Hong, L., Suh, B., Chi, E.H.: Tweets from Justin Bieber's heart: the dynamics of the location field in user profiles. In: Proceedings of the SIGCHI Conference on Human Factors in Computing Systems, pp. 237–246. ACM (2011)
25. Mahmud, J., Nichols, J., Drews, C.: Where is this tweet from? Inferring home locations of Twitter users. In: Proceedings of the Sixth International AAAI Conference on Weblogs and Social Media (ICWSM 2012), pp. 511–514 (2012)
26. Twitter Inc: Historical PowerTrack. http://support.gnip.com/apis/historicalapi/. Accessed 01 July 2018
27. Twitter Inc: Get Tweet timelines. https://developer.twitter.com/en/docs/tweets/timelines/api-reference/get-statuses-usertimeline.html. Accessed 01 July 2018
28. USGS: Water resources of the united states-national water information system (NWIS) mapper. https://maps.waterdata.usgs.gov/mapper/index.html. Accessed 04 Apr 2019
29. Schmidt, S., Achenbach, J., Somashekhar, S.: Puerto Rico entirely without power as Hurricane Maria hammers island with devastating force (2017)
30. Frailing, K., Wood Harper, D.: School kids and oil rigs: two more pieces of the Post-Katrina Puzzle in New Orleans. Am. J. Econ. Sociol. **69**(2), 717–735 (2010)
31. Frailing, K., Harper, D.W.: The Sociology of Katrina: Perspectives on a Modern Catastrophe. Rowman & Littlefield, Lanham (2010)
32. Frailing, K., Harper, D.W.: Crime and Criminal Justice in Disaster, 2nd edn. Carolina Academic Press, Durham (2012)
33. Frailing, K., Harper, D.W., Serpas, R.: Changes and challenges in crime and criminal justice after disaster. Am. Behav. Sci. **59**(10), 1278–1291 (2015)
34. Twitter: Twitter media policy (2018). Accessed 01 July 2018

Current Trends in Collaborative Filtering Recommendation Systems

Sana Abida Amin[1], James Philips[2], and Nasseh Tabrizi[2(✉)]

[1] Florida International University, Miami, FL 33199, USA
samin009@fiu.edu
[2] East Carolina University, Greenville, NC 27858, USA
philipsjl6@students.ecu.edu, tabrizim@ecu.edu

Abstract. Many different approaches for designing recommendation systems exist, including collaborative filtering, content-based, and hybrid approaches. Following an overview of different collaborative filtering recommendation system design methodologies, this paper reviews 71 journals articles and conference papers to provide a detailed literature review of model-based collaborative filtering. The articles selected for this review were published within the last decade between 2008–2018. They are classified by database, application field, methodology, and publication year. Papers using Clustering, Bayesian, Association Rule, Neural Networks, Regression, and Ensemble methodologies are surveyed. Application areas include books, music, movies, social networks, and business. This survey also analyzes the type of the data that was used for application field. This literature review identifies trends for model-based collaborative filtering and through empirical results gives insight into future research trajectories in this field.

Keywords: Collaborative filtering · Recommendation system · Methodologies · Applications

1 Introduction

Recommendation systems have become a ubiquitous part of everyday life. These systems are encountered daily in e-commerce sites such as Amazon and eBay, social media websites such as Instagram and Facebook, and entertainment websites including Netflix and Pandora. Yet, prior to their popularity in online e-commerce and entertainment, they have been an active research area since the mid-1990s [1]. Recommendation systems are a subset of "information filtering." They predict the rating user would give to an item. Using data-mining techniques and "prediction algorithms", these systems make highly relevant predictions for an active user [1]. Within an online context, this system helps users find products such as books, movies, or music by using ratings from the website's users to recommend relevant products based on the preferences of similar users [1].

This research is supported in part by grant #1560037 from the National Science Foundation.

Y. Xia and L.-J. Zhang (Eds.): SERVICES 2019, LNCS 11517, pp. 46–60, 2019.
https://doi.org/10.1007/978-3-030-23381-5_4

Recommendation systems are classified into Collaborative Filtering, Content-based Filtering, Cluster-based Filtering, and Hybrid approaches that combine item-based and user-based similarity [2]. Of these diverse approaches, Collaborative Filtering is the most well-known. There are two types of collaborative filtering "Model Based" and "Memory based" [2]. Collaborative filtering uses information filtering techniques to recommend items or products based on previous purchase history [2]. However, Collaborative filtering has some limitations such as cold start, sparsity, and scalability that cause poor performance. Over the years many methodologies have been developed to improve the performance of the system while avoiding the challenge of information overload. This literature review discusses the Model-Based Collaborative Filtering technique and current methodologies.

Moreover, this paper reviews academic journal articles and conference papers that were published between 2008–2018 that discuss the methodologies of Model-Based Collaborative filtering recommendation systems. These papers are classified by application field, methodologies, and publication year to illuminate research trajectories in the field of model-based collaborative filtering and its significance for Big Data applications.

2   Background

Recommendation systems are software tools that make predictions for users based on their likes and dislikes. Recommender systems can be divided into two groups Personalized and Non-Personalized [1, 3]. Personalized recommendation systems recommend products or items based on individual criteria. In contrast, Non-Personalized Recommendation systems do not depend on individual criteria [3]. Instead, this type of system provides the same recommendation to all the users regardless of personal likes, dislikes, or demographic location.

Moreover, Recommendation systems are classified into collaborative filtering, content based and hybrid [1]. Collaborative filtering is a technique that makes a recommendation by finding users that have similar interests [2]. Content-based systems make recommendation by using users' background information from interaction with the system such as browsing history [2]. However, hybrid systems make prediction by integrating techniques of collaborative and content-based together [2]. However, with the challenge of recommendation systems in the Big Data context, it is more difficult to provide an accurate recommendation to an active user due to the overwhelming size of the data.

To overcome this exponential increase of Big Data challenges, researchers have been working on different techniques that can be applied to address the solution for collaborative-filtering based recommendation systems. According to [2], other challenges to successful recommendation are lack of data, data evolution, and variety of users' preferences. Furthermore, other limitations of collaborative filtering are cold start, scalability, and complexity [1, 2]. Researchers are actively researching different techniques to eliminate these challenges and limitations.

3 Overview of Collaborative Filtering and Its Techniques

In recent years, extensive work has been done on recommendation methodologies and approaches. This section reviews general concepts and algorithms. Recommendation Systems are generally classified into content-based systems, collaborative-filtering-based systems, and hybrid designs that combine these two techniques in a single system. However, this review will emphasize those design principles relevant to Collaborative-Filtering.

3.1 Collaborative Filtering

Collaborative Filtering is one of the most popular recommendation system techniques. This system calculates similarities between the users of the system and predicts items based on their similar patterns. This method of recommendation system uses data ratings for an item provided by an active user from a large database known as a "user-item matrix" [4]. It then calculates the similarities between users by matching interest and preference for an item; this process is known as "neighborhood" [4]. Users will sometimes receive a recommendation for an item which they did not rate; however, since users with similar interests are grouped into clusters, the item was already rated by users within the same neighborhood as the original user.

3.2 Collaborative Filtering Process

The main concern of collaborative filtering is to recommend items for a user based on existing ratings within the neighborhood of similar users. This process involves the following steps:

1. The system represents the entire space as a two-dimensional matrix. Within a two-dimensional matrix R, where i and j correspond to users and items, respectively. Each rating given by the system is the value Rij. If an item has not been rated by a user, its item has value 0.
2. This matrix R is then used to predict the rating for item j provided by user i in order to make recommendations for a list of N items that the user might like.

3.3 Collaborative Filtering Algorithms and Methods

Two types of Collaborative Filtering methods exist, memory-based and model-based [5, 17].

Memory-Based Techniques. Memory-based Collaborative Filtering techniques use the system's memory to produce predictions. These techniques find previous users that have identical or similar interests as current users as a basis to predict items for current users. Once the current user rates items from its user-item dataset, algorithms use these ratings to combine users' interests to make new predictions. Memory-based systems can be further divided into user-based and item-based methods [6].

User-Based Method. This method compares the users' rating pattern for an item. Then based on the comparison, it makes prediction of an item for a user which has been rated by another user within the neighborhood. This technique calculates the prediction based on user similarity by comparing their evaluations on same item [6].

Item-Based Method. This method computes the prediction based on the similarity between items. The equivalence between the user-based and item-based recommendations can be calculated using the Pearson correlation coefficient and cosine similarity [6].

Model-Based Techniques. Model-Based recommendation systems incorporate machine-learning and use user ratings and preferences to learn a model about the user. Hidden characteristics and item preferences train the model to offer new predictions for a user [7]. This system also uses users' implicit information, such as music played, books read, or websites visited to make recommendation to the users. Methods for Model-Based Collaborative Filtering include Cluster Models, Bayesian Networks, Association Rule, and Neural Networks [2]. Ensemble techniques combine one or more of these approaches. Hybrid approaches combine collaborative filtering with another technique. Figure 1 shows various model-based techniques for collaborative filtering.

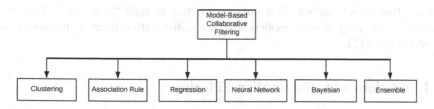

Fig. 1. Model-based collaborative filtering techniques

Clustering. Clustering algorithms are a form of unsupervised machine learning that structure data based on a predefined model. Used extensively in image processing, pattern recognition, and statistical parameter models, this model partitions data into clusters. Once stable clusters have been created, parameters will be used to average out the recommendation [4]. In this approach users with similar interests are grouped together to make recommendations within the neighborhood [8]. A good clustering method possesses high intra-cluster similarity and low inter-cluster similarity [4]. The most commonly used clustering algorithm is K-mean. This algorithm is relatively simple to implement and consistently provides better accuracy for recommendations than other algorithms.

Bayesian Networks. Bayesian networks are based on conditional probability and Bayes theorem [4]. Ratings will be determined from each node which represent the item.

Association Rule Mining. Association Rule Mining is used to describe the relationship between items that have been purchased together. This mining algorithm makes prediction about an item, based on items that have been purchased in the past or purchased concurrently in a transaction [9].

Neural Network. Neural Networks resemble the human brain where many neurons link to each other. In a neural network, neurons are connected in layers that includes input and output nodes. According to [10] there are many kinds of neural network, but one basic neural network is Multiple Layer Perceptron (MLP). There are two kinds of neural networks commonly used in collaborative filtering recommendation systems, user-based and item-based [10]. In user-based networks, input nodes refer to the user's previous preferences and output nodes refer to user's preference for an item [10]. However, item-based input node refers to an item preference and output node corresponds to a user's preference for the target item [10].

Regression. Regression is the statistical process to determine the relationship within datasets. Specifically, it is a method to ascertain the association between independent variable(s) and a dependent variable [5]. The main purpose for this approach is to determine users' rating within their neighborhood [11]. This approach helps predict hidden characteristics of relationships among users' rating "habits" [11]. This method calculates users' ratings and identifies common patterns between users and their neighborhood.

Ensemble. An ensemble model is the combination of two or more algorithms and methods to provide recommendations in order to improve the results from a single method. This model contains four components that include "Boosting", "Bagging," "Fusion" (combining several models that use collaborative filtering methods), and "Randomness" [12].

4   Limitations of Collaborative Filtering

Although collaborative filtering is among the most popular recommendation system techniques, there are some challenges discussed below.

4.1   Cold Start

Cold start can be described in three scenarios, such as "new community", "new item" and "new user" [13].

New Community. This problem refers to the challenge of gathering sufficient initial ratings to make recommendations to a new group of users [13].

New Item. This problem occurs when a new item enters the system. In the beginning, new items do not have ratings; therefore, they are less likely to be recommended. These new items go unnoticed by large parts of the community [13].

New User. New user is a significant challenge in collaborative filtering recommendation systems because there is no history of preferences to use as a basis for recommendations [13]. The user's preferences are completely unknown to the system. Therefore, the system is unable to make a reliable recommendation to the user.

4.2 Scalability

Collaborative filtering uses billions of data to make reliable recommendation to the users which requires extensive computation resources. As collaborative filtering information grows exponentially, processing becomes expensive and inaccurate ratings can result from this Big Data challenge [2].

4.3 Sparsity

Recommendation systems rely on a massive catalog of item rankings. However, only a subset of this data is used for individual items, which leads some items having few ratings. This sparsity of rankings makes it difficult for a system to calculate recommendations [2].

Because of these limitations of collaborative filtering, one objective of this literature review is to emphasize articles and papers in conference proceedings that are dedicated to alleviating these problems using machine learning methods.

5 Research Methodology for Literature Review

It is necessary to establish the research methodology for this review and criteria for inclusion of papers. The different articles and conference proceedings are surveyed based on distributions by year, methodology, application field and dataset. This paper's chronological range was for articles and conference proceedings published in the last decade from 2008 to 2018. Moreover, this study only includes those articles that conducted an experiment using dataset with one or more methodologies. The following databases were used in this study: IEEE Xplore, Science Direct, Springer Link, and ACM Digital Library.

Figure 2 shows the distribution of research articles by database.

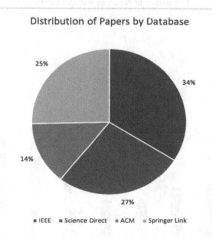

Fig. 2. Distribution of articles by database

Thirty-four percent of the articles, or 24 of 71, are from IEEE. Twenty-seven percent of the articles are from Science Direct, which is 19 out of 71 articles. Twenty-five percent of the articles came from Springer Link, or 18 of 71 articles. Finally, fourteen percent of the articles are from the ACM digital library, which is 10 of 71 articles.

The key selection criteria for papers in this study were collaborative filtering recommendation system articles that were performing experiments on real data sets and trying to alleviate the inherent challenges for collaborative filtering techniques discussed above. Moreover, this study was confined to literature published in English. The keywords chosen for search process were: "recommendation system", "recommender system", "methodologies and recommendation system", "Association rule and recommendation system", "regression and recommendation system", "neural network and recommendation system", and "ensemble and recommendation system." Although consulted, survey papers were not included in this study unless they included an experiment on a dataset.

6 Classification of Papers for Literature Review

Seventy-one articles on methodologies for collaborative-filtering recommendation system were reviewed in this study. These papers were classified into categories by year, application field, and methodology. This section discusses the detailed classification of the articles and conference papers surveyed for this literature review.

6.1 Distribution of Papers by Year

The research papers were selected by year of publication between 2008–2018 as shown in Fig. 3. The majority of research papers published between 2016–2018 were using different methods to implement in different application areas such as books, music, social networks, travel, and education to find solutions for the limitations of collaborative filtering discussed in Sect. 4 above.

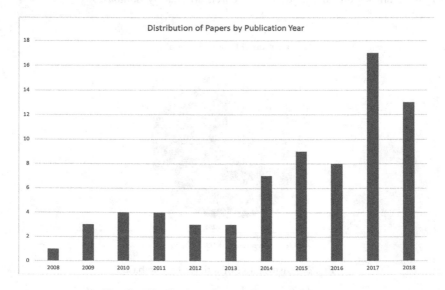

Fig. 3. Distribution of papers by publication year

6.2   Distribution of Papers by Methodology/Technique

Distribution of papers by different techniques for collaborative filtering is shown below in Fig. 4. From the 71 papers that were reviewed, the clustering technique predominates [14, 16, 24, 34, 41, 42, 51, 52, 54, 70, 74, 75, 78, 79, 82]. This technique has proven beneficial to eliminate the challenges in collaborative filtering. Association rule was researched in the earlier years of the period considered for this survey [21, 29, 36, 39, 40, 46, 64]. However, more recently, researchers have become more interested in implementing the association technique alongside another method within a hybrid or ensemble approach [31, 37, 38, 44, 47, 55, 58, 65, 68, 80]. Bayesian prediction ranks next in number of papers. In the majority of cases Bayesian technique has been used to implement Movie, social media and location-based recommendation systems [18, 19, 22, 28, 44, 45, 49, 50, 53, 57, 60, 62, 63, 80]. The use of neural networks in recommender systems is similar to the Bayesian [20, 23, 26, 32, 37, 47, 58, 66, 71, 72, 76]. Finally, regression has received the least study of these methodologies [11, 61, 73, 77, 81]. However, increased research on ensemble techniques may have decreased the research devoted exclusively to regression. As with Association Rule methodology, researchers are interested in implementing regression with other techniques to improve recommendation system performance.

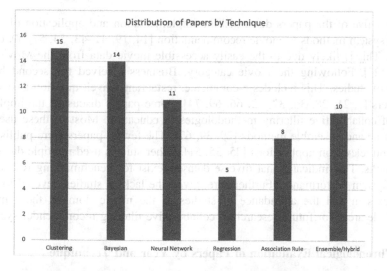

Fig. 4. Distribution of papers by technique/methodology

6.3   Distribution of Papers by Application

The distribution of papers by application is represented in Fig. 5.

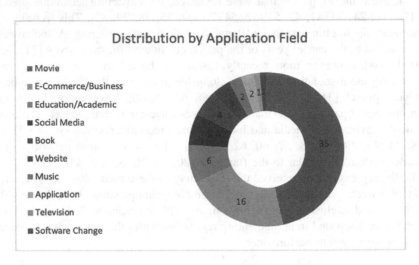

Fig. 5. Distribution by application field

Thirty-five of the papers discussed the implementation and application of recommender system methods to movie recommendation [14, 19, 41–43, 54, 57, 60, 61, 71, 78, 81]. This is likely due to the easily accessible movie data from the Movie Lens dataset [27]. Following the movie category, Business received the second highest number of studies. This includes e-commerce, restaurant, travel, question answering , and others [8, 21, 28, 36, 37, 52, 66, 69, 74]. Some papers discussed the implementation of collaborative filtering methodologies in education. Most of these use association rule and ensemble methods [29, 39, 67]. The fewest papers were published in music and television applications [25, 35, 51]. Other studies used multiple datasets as benchmarks. This indicates that diverse datasets exist for benchmarking Recommendation System performance. Furthermore, given the lack of studies devoted to music and television and the abundance of studies in the movie domain, these may be worthwhile areas for future research in collaborative filtering recommender systems.

6.4   Chronological Evaluation of Papers by Year and Technique

During the late 2000s Association Rule was predominant [21, 29, 30, 40, 46, 55, 64, 68]. Following a hiatus, researchers resumed interest in this method in 2017. Clustering can be considered the most stable research category because it has been consistently investigated since 2009 [14, 16, 24, 28, 34, 42, 51, 52, 54, 70, 74, 75, 78, 79, 82]. Few papers in the early period of this study examined this method. However, more recent studies in clustering have focused on implementing fuzzy clustering instead of traditional clustering [41]. Another stable research category is Bayesian prediction with papers consistently appearing on this technique with almost yearly regularity [18, 19, 22, 28, 44, 45, 49, 50, 53, 57, 60, 62, 63]. Few papers from 2009–2010 and 2014–2017 discuss the Regression technique [11, 33, 59, 61, 73, 77, 81]. Low frequency of articles for a given technique could indicate that researchers are transitioning to a new research emphasis, such as such as ensemble methods. Only one paper from 2009 discussed ensemble technique; but, since 2016, this method has grown in popularity among researchers [12, 15, 31, 37, 58, 65, 80]. The ascendant popularity of Ensemble techniques indicates a current trajectory of recommender system research. Moreover, many researchers are interested in implementing this method for different application fields such as context-based, location-based, demographics, education, and social media [6, 28–30, 39, 44, 46, 48, 52, 63, 66, 68, 75].

7 Conclusion

Recommendation systems predict users rating of products or items based on users' likes and dislikes. These systems utilize users' background history and current information by using implicit features and explicit feedback to provide high quality, relevant and diverse recommendations. This paper discussed model-based collaborative filtering, examined its different methodologies and applications, and surveyed the results of a literature review of model-based collaborative filtering. For this review, 71 articles and conference papers published between 2008–2018 were surveyed to furnish insight about the trajectories in the field of Big Data and collaborative filtering techniques and applications.

Papers were classified in terms of database, publication year, methodologies or technique, application field, and technique and publication year. There are few papers using Ensemble method in early years covered in this survey. However, more papers using Ensemble models have been published within the past few years. Moreover, it was also observed that generally fewer papers were published between 2008–2013 on collaborative filtering, then after 2014 the publication count rises, and the trend reaches its peak in 2017.

For application field, there were more papers implementing collaborative filtering methodologies on movie datasets than any other type of dataset. This is due to the wide availability of movie datasets such as MovieLens and Netflix [27, 56]. As a future goal for research, it is important to investigate further ensemble methods and the potential limitations of this method. Also, it is important to expand the availability and variety of research datasets for collaborative filtering recommendation systems. With the ascendancy of Big Data, it is evident that Recommender Systems will continue to be a crucial tool to navigate the overwhelming variety choices across numerous domains.

References

1. Shah, K., Salunke, A., Dongare, S., Antala, K.: Recommender systems: an overview of different approaches to recommendations. In: International Conference on Innovations in Information, Embedded and Communication Systems, pp. 1–4. IEEE (2017)
2. Sharma, M., Mann, S.: A survey of recommender systems: approaches and limitations. Int. J. Innov. Eng. Technol. 2(2), 8–14 (2013)

3. Aditya, P., Budi, I., Munajat, Q.: A comparative analysis of memory-based and model-based collaborative filtering on the implementation of recommender system for e-commerce in Indonesia: a case study PT X. In: International Conference on Advanced Computer Science and Information Systems, pp. 303–308. IEEE (2016)

4. Khatwani, S., Chandak, M.: Building personalized and non-personalized recommendation systems. In: International Conference on Automatic Control and Dynamic Optimization Techniques, pp. 623–628. IEEE (2016)

5. Isinkaye, F., Folajimi, Y., Ojokoh, B.: Recommendation systems: principles, methods, and evaluation. Egyptian Inform. J. **16**(3), 261–273 (2015)

6. Pareek, J., Jhaveri, M., Kapasi, A., Trivedi, M.: SNetRS: social networking in recommendation system. In: Meghanathan, N., Nagamalai, D. (eds.) Advances in Computing and Information Technology. Advances in Intelligent Systems and Computing, vol. 177. Springer, Heidelberg (2013). https://doi.org/10.1007/978-3-642-31552-7_21

7. Liphoto, M., Du, C., Ngwira, S.: A survey on recommender systems. In: International Conference on Advances in Computing and Communication Engineering, pp. 276–280. IEEE (2016)

8. Kim, K., Ahn, H.: A recommender system using GA k-means clustering in an online shopping market. Expert Syst. Appl. **34**(2), 1200–1209 (2008)

9. Sun, X., Kong, F., Chen, H.: Using quantitative association rules in collaborative filtering. In: Fan, W., Wu, Z., Yang, J. (eds.) WAIM 2005. LNCS, vol. 3739, pp. 822–827. Springer, Heidelberg (2005). https://doi.org/10.1007/11563952_87

10. Kim, E., Kim, M., Ryu, J.: Collaborative filtering based on neural networks using similarity. In: Wang, J., Liao, X.F., Yi, Z. (eds.) ISNN 2005. LNCS, vol. 3498, pp. 355–360. Springer, Heidelberg (2005). https://doi.org/10.1007/11427469_57

11. Ge, X., Liu, J., Qi, Q., Chen, Z.: A new prediction approach based on linear regression for collaborative filtering. In: Eighth International Conference on Fuzzy Systems and Knowledge Discovery, pp. 2586–2590. IEEE (2011)

12. Bar, A., Rokach, L., Shani, G., Shapira, B., Schclar, A.: Boosting simple collaborative filtering models using ensemble methods. Arkiv, 21 pp. (2012)

13. Bobadilla, J., Ortega, F., Hernando, A., Gutiérrez, A.: Recommender systems survey. Knowl.-Based Syst. **46**, 109–132 (2013)

14. Ahmed, M., Imtiaz, M., Khan, R.: Movie recommendation system using clustering and pattern recognition network. In: 8th Annual Computing and Communication Workshop and Conference, pp. 143–147. IEEE (2018)

15. Ayaki, T., Yanagimoto, H., Yoshioka, M.: Recommendation from access logs with ensemble learning. Artif. Life Robot. **22**(2), 163–167 (2017)

16. Aytekin, T., Karakaya, M.: Clustering-based diversity improvement in top-n recommendation. J. Intell. Inf. Syst. **42**(1), 1–18 (2014)

17. Bai, T., Wen, J., Zhang, J., Zhao, W.: A neural collaborative filtering model with interaction-based neighborhood. In: Proceedings of the 2017 ACM Conference on Information and Knowledge Management, pp. 1979–1982. ACM (2017)

18. Barbieri, N., Costa, G., Manco, G., Ortale, R.: Modeling item selection and relevance for accurate recommendations: a Bayesian approach. In: Proceedings of the 5th ACM Conference on Recommender Systems, pp. 21–28. ACM (2011)

19. Beutel, A., Murray, K., Faloutsos, C., Smola, A.: Cobafi: collaborative Bayesian filtering. In: Proceedings of the 23rd International Conference on the World Wide Web, pp. 97–108. ACM (2014)

20. Bi, X., Jin, W.: An improved collaborative filtering similarity model based on neural networks. In: International Conference on Intelligent Transportation, Big Data, and Smart Cities, pp. 85–89. IEEE (2015)

21. Cakir, O., Aras, M.: A recommendation engine by using association rules. Procedia Soc. Behav. Sci. **62**, 452–456 (2012)
22. Chatzis, S.: Nonparametric Bayesian multitask collaborative filtering. In: Proceedings of the 22nd ACM International Conference on Information and Knowledge Management, pp. 2149–2158. ACM (2013)
23. Chen, T., Sun, Y., Shi, Y., Hong, L.: On sampling strategies for neural network-based collaborative filtering. In: Proceedings of the 23rd ACM SIGKDD International Conference on Knowledge Discovery and Data Mining, pp. 767–776. ACM (2017)
24. Tsai, C., Hung, C.: Cluster ensembles in collaborative filtering recommendation. Appl. Soft Comput. **12**(4), 1417–1425 (2012)
25. Cho, Y., Moon, S., Jeong, S.: Learning listener's preference for music recommender system. In: Proceedings of the 2015 International Conference on Big Data Applications and Services, pp. 229–232. ACM (2015)
26. Ebesu, T., Fang, Y.: Neural semantic personalized ranking for item cold-start recommendation. Inf. Retrieval J. **20**(2), 109–131 (2017)
27. Harper, F., Konstan, J.: The movielens datasets: history and context. ACM Trans. Interactive Intell. Syst. **5**(4), 19 (2016)
28. Gao, S., Guo, G., Lin, Y., Zhang, X., Liu, Y., Wang, Z.: Pairwise preference over mixed-type item-sets based Bayesian personalized ranking for collaborative filtering. In: 15th International Conference on Pervasive Intelligence and Computing, pp. 30–37. IEEE (2017)
29. Garcia, E., Romero, C., Ventura, S., De Castro, C.: A collaborative educational association rule mining tool. Internet High. Educ. **14**(2), 99–132 (2009)
30. Garcia, E., Romero, C., Ventura, S., De Castro, C.: An architecture for making recommendations to courseware authors using association rule mining and collaborative filtering. User Modeling User-Adapted Interact. **19**, 99–132 (2009)
31. Gong, S., Ye, H., Tan, H.: Combining memory-based and model-based collaborative filtering in recommender system. In: Pacific-Asia Conference on Circuits, Communications, and Systems. IEEE (2009)
32. He, X., Liao, L., Zhang, H., Nie, L., Hu, X., Chua, T.: Neural collaborative filtering. In: Proceedings of the 26th International Conference on World Wide Web, pp. 173–182. ACM (2017)
33. Hwang, C., Kao, Y., Yu, P.: Integrating multiple linear regression and multicriteria collaborative filtering for better recommendation. In: 2010 International Conference on Computational Aspects of Social Networks, pp. 229–232. IEEE (2010)
34. Javari, A., Jalili, M.: Cluster-based collaborative filtering for sign prediction in social networks with positive and negative links. ACM Trans. Intell. Syst. Technol. **5**(2), 24 (2014)
35. Jiang, M., Yang, Z., Zhao, C.: What to play next? A RNN-based music recommendation system. In: 51st Asilmar Conference on Signals, Systems, and Computers, pp. 356–358. IEEE (2017)
36. Jooa, J., Bangb, S., Parka, G.: Implementation of recommendation system using association rules and collaborative filtering. Procedia Comput. Sci. **91**, 944–952 (2016)
37. Paradarami, T., Bastian, N., Wightman, J.: A hybrid recommender system using artificial neural networks. Expert Syst. Appl. **83**, 300–313 (2017)
38. Kant, S., Mahara, T.: Nearest bi-clusters collaborative filtering framework with fusion. J. Comput. Sci. **25**, 204–212 (2017)
39. Kazienko, P., Pilarczyk, M.: Hyperlink recommendation based on positive and negative association rules. New Gen. Comput. **26**(3), 227–244 (2008)
40. Kiran, R., Kitsuregawa, M.: An improved neighborhood-restricted association rule-based recommender system. In: Proceedings of the 24th Australasian Database Conference, pp. 43–50 (2013)

41. Koohi, H., Kiani, K.: User based collaborative filtering using fuzzy c-means. Measurement **91**, 134–139 (2016)
42. Li, J., et al.: Category preferred canopy-k-means based collaborative filtering algorithm. Future Gen. Comput. Syst. **93**, 1046–1054 (2018)
43. Li, W., Li., X., Yao, M., Jiang, J., Jin, Q.: Personalized fitting recommendation based on support vector regression. Hum.-Centric Comput. Inf. Sci. **5**(1), 21 (2015)
44. Lin, K., Wang, J., Zhang, Z., Chen, Y., Xu, Z.: Adaptive location recommendation algorithm based on location-based social networks. In: 10th International Conference on Computer Science and Education, pp. 137–142. IEEE (2015)
45. Liu, C., Jin, T., Hoi, S., Zhao, P., Sun, J.: Collaborative topic regression for online recommender systems: an online and Bayesian approach. Mach. Learn. **106**(5), 651–670 (2017)
46. Liu., Y.: Data mining of university library management based on improved collaborative filtering association rules algorithm. Wirel. Pers. Commun. **102**, 3781–3790 (2018)
47. Liu, Y., Wang, S., Khan, M., He, J.: A novel deep hybrid recommender system based on auto-encoder with neural collaborative filtering. Big Data Mining Anal. **1**(3), 211–221 (2018)
48. Logesh, R., Subramaniyaswamy, V., Vijayakumar, V., Gao, X., Indragandhi, V.: A hybrid quantum-induced warm intelligence clustering for the urban trip recommendation in smart city. Future Gen. Comput. Syst. **83**, 653–673 (2017)
49. Lopes, R., Assunção, R., Santos, R.: Efficient Bayesian methods for graph-based recommendation. In: Proceedings of the 10th ACM Conference on Recommender Systems, pp. 333–340. ACM (2016)
50. Luo, C., Cai, X.: Bayesian wishart matrix factorization. Data Mining Knowl. Discov. **30**(5), 1166–1191 (2016)
51. Ma, Z., Yang, Y., Wang, F., Li, C., Li, L.: The SOM based improved k-means clustering collaborative filtering algorithm in TV recommendation system. In: Second International Conference on Advanced Cloud and Big Data, pp. 288–295 (2014)
52. Margaris, D., Georgiadis, P., Vassilakis, C.: A collaborative filtering algorithm with clustering for personalized web service selection in business processes. In: 9th International Conference on Research Challenges in Information Science, pp. 169–180 (2015)
53. Maurya, A., Telang, R.: Bayesian multi-view models for member-job matching and personalized skill recommendations. In: IEEE International Conference on Big Data, pp. 1193–1202. IEEE (2017)
54. Mittal, N., Nayak, R., Govil, M., Jain, K.: Recommender system framework using clustering and collaborative filtering. In: 3rd International Conference on Emerging Trends in Engineering and Technology, pp. 555–558. IEEE (2010)
55. Nagarnaik, P., Thomas, A.: Survey on recommendation system methods. In: 2nd International Conference on Electronics and Communication Systems, pp. 1496–1501. IEEE (2015)
56. Bennett, J., Lanning, S.: The netflix prize. In: Proceedings of KDD Cup and Workshop, pp. 75–79. ACM (2007)
57. Nguyen, L.: A new approach for collaborative filtering based on Bayesian network inference. In: 7th International Conference on Knowledge Discovery, Knowledge Engineering and Knowledge Management, pp. 475–480. IEEE (2015)
58. Nilashi, M., Bagherifard, K., Rahmani, M., Rafe, V.: A recommender system for tourism industry using cluster ensemble and prediction machine learning techniques. Comput. Ind. Eng. **109**, 357–368 (2017)
59. Nilashi, M., Jannach, D., bin Ibrahim, O., Ithnin, N.: Clustering and regression-based multi-criteria collaborative filtering with incremental updates. Inf. Sci. **293**, 235–250 (2015)

60. Pan, W., Chen, L.: Group Bayesian personalized ranking with rich interactions for one-class collaborative filtering. Neurocomputing **207**, 501–510 (2016)
61. Park, S., Chu, W.: Pairwise preference regression for cold-start recommendation. In: Proceedings of the 3rd ACM Conference on Recommender Systems, pp. 21–28. ACM (2009)
62. Qiu, H., Liu, Y., Guo, G., Sun, Z., Zhang, J., Nguyen, H.: BPRH: Bayesian personalized ranking for heterogeneous implicit feedback. Inf. Sci. **453**, 80–98 (2018)
63. Rho, W., Cho, S.: Context-aware smartphone application category recommender system with modularized Bayesian networks. In: 10th International Conference on Natural Computation, pp. 775–779. IEEE (2014)
64. Rolfsnes, T., Moonen, L., Di Alesio, S., Behjati, R., Binkley, D.: Aggregating association rules to improve change recommendation. Empirical Softw. Eng. **23**(2), 987–1035 (2018)
65. Schclar, A., Tsikinovsky, A., Rokach, L., Meisels, A., Antwarg, L.: Ensemble methods for improving the performance of neighborhood-based collaborative filtering. In: Proceedings of the 3rd ACM Conference on Recommender Systems, pp. 261–264. ACM (2009)
66. Sohrabi, B., Mahmoudian, P., Raeesi, I.: A framework for improving e-commerce websites usability using a hybrid genetic algorithm and neural network system. Neural Comput. Appl. **21**(5), 1017–1029 (2012)
67. Thai-Nghe, N., Drumond, L., Krohn-Grimberghe, A., Schmidt-Thieme, L.: Recommender system for predicting student performance. Procedia Comput. Sci. **1**(2), 2811–2819 (2010)
68. Viktoratos, I., Tsadiras, A., Bassiliades, N.: Combining community-based knowledge with association rule mining to alleviate the cold start problem in context-aware recommender systems. Expert Syst. Appl. **101**, 78–90 (2018)
69. Wang, J., Sun, J., Lin, H., Dong, H., Zhang, S.: Convolutional neural networks for expert recommendation in community question answering. Sci. China Inf. Serv. **60**(11), 110–120 (2017)
70. Wang, S., Zhao, Z., Hong, X.: The research on collaborative filtering recommendation algorithm based on improved clustering processing. In: IEEE International Conference on Computer and Information Technology; Ubiquitous Computing and Communications; Dependable, Autonomic and Secure Computing; Pervasive Intelligence and Computing, pp. 1012–1015. IEEE (2015)
71. Wei, J., He, J., Chen, K., Zhou, Y., Tang, Z.: Collaborative filtering and deep learning based recommendation system for cold start items. Expert Syst. Appl. **69**, 29–39 (2017)
72. Wu, H., Zhang, Z., Yue, K., Zhang, B., He, J., Sun, L.: Dual-regularized matrix factorization with deep neural networks for recommender systems. Knowl.-Based Syst. **145**, 46–58 (2018)
73. Wu, J., Miao, Z.: Regression-based fusion prediction for collaborative filtering. In: International Conference on Cloud Computing and Big Data, pp. 312–319. IEEE (2013)
74. Xiaojun, L.: An Improved clustering-based collaborative filtering recommendation algorithm. Cluster Comput. **20**(2), 1281–1288 (2017)
75. Xie, X., Wang, B.: Web Page recommendation via twofold clustering: considering user behavior and topic relation. Neural Comput. Appl. **29**(1), 235–243 (2018)
76. Xu, Y., Zhu, Y., Shen, Y., Yu, J.: Leveraging app usage contexts for app recommendation: a neural approach. World Wide Web 1–25 (2018)
77. Yang, M., Li, Y., Zhang, Z.: Scientific articles recommendation with topic regression and relational matrix factorization. J. Zhejiang Univ. Sci. C **15**(11), 984–998 (2014)
78. Zarzour, H., Al-Sharif, Z., Al-Ayyoub, M., Jararweh, Y.: A new collaborative filtering recommendation algorithm based on dimensionality reduction and clustering techniques. In: 9th International Conference on Information and Communication Systems, pp. 102–106. IEEE (2018)

79. Zhang, C., Dai, J., Li, P., Li, Q., Luo, X.: Two-phase clustering-based collaborative filtering algorithm. In: 5th International Conference on Management of e-Commerce and e-Government, pp. 19–23. IEEE (2011)
80. Zhang, H., Ganchev, I., Nikolov, N., Ji, Z., Odroma, M.: Hybrid recommendation for sparse rating matrix: a heterogeneous information network approach. In: Proceedings of the IEEE Advanced Information Technology, Electronic and Automation Control Conference, pp. 740–744. IEEE (2017)
81. Zhang, H., Min, F., Shi, B.: Regression-based three-way recommendation. Inf. Sci. **378**, 444–461 (2017)
82. Zhao, W., Zhang, H.: The improved item-based clustering collaborative filtering algorithm based on Hadoop (2017)

Big Data Quality: A Data Quality Profiling Model

Ikbal Taleb[1] , Mohamed Adel Serhani[2](✉) ,
and Rachida Dssouli[1]

[1] Concordia University, Montreal, QC H3G 2W1, Canada
i_taleb@live.concordia.ca, rachida.dssouli@concordia.ca
[2] UAE University, Al Ain, Abu Dhabi, United Arab Emirates
serhanim@uaeu.ac.ae

Abstract. Big Data is becoming a standard data model, and it is gaining wide
adoption in the digital universe. Estimating the Quality of Big Data is recognized to be essential for data management and data governance. To ensure a fast
and efficient data quality assessment represented by its dimensions, we need to
extend the data profiling model to incorporate also quality profiling. The latter
encompasses more value-added quality processes that go beyond data and its
corresponding metadata. In this paper, we propose a Data Quality Profiling
Model (BDQPM) for Big Data that involves several modules such as sampling,
profiling, exploratory quality profiling, quality profile repository (QPREPO),
and the data quality profile (DQP). Thus, the QPREPO plays an important role
in managing many quality-related elements such as data quality dimensions and
their related metrics, pre-defined quality actions scenarios, pre-processing
activities (PPA), their related functions (PPAF), and the data quality profile. Our
exploratory quality profiling method discovers a set of PPAF from systematic
predefined quality actions scenarios to leverage the quality trends of any data set
and show the cause and effects of such a process on the data. Such a quality
overview is considered as a preliminary quality profile of the data. We conducted a series of experiments to test different features of the BDQPM including
sampling and profiling, quality evaluation, and exploratory quality profiling for
Big Data quality enhancement. The results prove that quality profiling tracks
quality at the earlier stage of Big data life cycle leading to quality improvement
and enforcement insights from exploratory quality profiling methodology.

Keywords: Big Data quality · Data Quality Profile · Profile repository ·
Data quality profiling

1 Introduction and Background

The big data ecosystem is defined as the way we gather, store, manipulate, analyzes
and get insight from a fast-increasing heterogeneous data. According to IBM [1], every
day huge amounts of data are generated; this data represents 2.5 quintillion bytes
(Exabyte (EB) = 10^{18} bytes) [2]. In 2000, 800,000 Petabyte (1 PB = 10^{15} bytes) of
data were stored. In 2020, the worldwide storage will reach 35 Zettabytes (1 ZB = 10^{21}

© Springer Nature Switzerland AG 2019
Y. Xia and L.-J. Zhang (Eds.): SERVICES 2019, LNCS 11517, pp. 61–77, 2019.
https://doi.org/10.1007/978-3-030-23381-5_5

bytes = 1 Trillion gigabytes). Which urge the need to automatically profile, characterize and categorize the quality of such data. These classifications are strongly coupled with the semantic meaning of what the data represents. In many cases, the data comes in a format and a quality state in which it is impossible to process immediately as it is, and if so, the results cannot guarantee a trustable analysis and insights. The importance of estimating and profiling the quality of Big Data is paramount and has priority over the other Big Data stages.

In this paper, we introduce a data quality profiling model (BDQPM) for Big Data that acts as a preliminary quality discovery in the Big data lifecycle before engaging with any analytics of a data source. Therefore, quality profiling will intervene before the pre-processing stage of Big Data. The exploratory quality profiling module is considered as the core of our model in which a pre-defined pre-processing quality scenarios actions are applied on data samples. These scenarios target specific Data Quality Dimensions (DQD's) while variating the DQD acceptance ratio level set (from min to max ratio). The resulted pre-processed samples will have their DQD evaluated in the quality evaluation module while the max ratio is not reached. Once the quality results are aggregated, an analysis is done and the set of pre-processing activity functions (PPAF's) that affected the DQD ratio are selected to be used as quality enhancements rules.

Our model relies on a quality profile repository (QPRepo) that handles all the related quality tables, from Pre-Processing Activities (PPA), Pre-Processing Activity Functions (PPAF), DQD's and metrics. The most important is the Data Quality Profile (DQP) that plays the role of a record book of quality that tracks all the data and quality profiling results, metadata, the pre-defined quality scenarios, the quality scores, and the quality reports. Finally, a set of experimentations to validate out model modules: (a) sampling and profiling, (b) sampling and quality evaluation and (c) exploratory quality profiling for Big Data quality improvement PPAF extraction.

The rest of paper is organized as follows: next section introduces Big Data and data quality fundamentals, definition, characteristics, and lifecycle. Section 3 surveys the most important research on quality profiling for Big Data. Section 4 introduces our Big Data quality profiling approach. Section 5 presents our Big Data quality profile repository. Section 6, analyzes and discusses our experimentations. Finally, the last section concludes the paper and highlights some ongoing and challenging research directions.

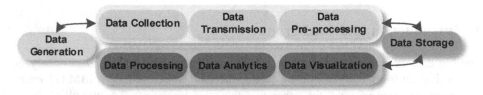

Fig. 1. Big Data key processes

1.1   Big Data Value Chain

As illustrated in Fig. 1, Big Data is handled through a lifecycle (also called value chain) that tracks and handles the data from its inception to insights generation. There are many stages where the data goes through to achieve a specific goal. In the following, a brief description of these stages:

- **Data Generation:** this is considered as the inception stage about the data sources, where the data is being engendered.
- **Data Acquisition:** it involves data collection, transmission and pre-processing. In the pre-processing, the data might be combined from different sources and tailored into a pre-defined format for the purpose of processing.
- **Data Processing and Analytics:** it consists of processing data using several analytics approaches and tools (e.g. Data Mining techniques, Machine Learning algorithms, Deep Learning).
- **Data Transmission and Storage:** consist of transmission of huge data over a network and the distribution and replication of storage.

Big Data Characteristics (V's). Big Data is also described by its characteristics that brand the Data as "***Big Data***". The initial Big Data characteristics are volume, velocity, variety [3, 8, 19, 22]. However, these got extended to cover other 9 extra Big data characteristics that we compiled with the illustrated Fig. 2. Moreover, V's represents the key elements affecting traditional data to become Big Data.

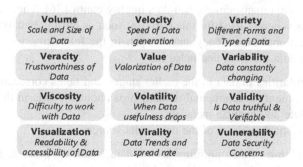

Fig. 2. Big Data characteristics

1.2   Data Quality

According to [18], data quality is not easy to define, its meanings are data domain dependent and context-aware. Data quality is continuously related to the quality of its data source [15]. It is also defined by its quality dimensions, metrics and assessments.

Data Quality Dimensions: To measure and manage data quality the concepts of a Data Quality Dimension (DQD) is presented in [4, 9, 20]. There are many quality dimensions classified under categories that define them. Some essential DQD categories are (a) the contextual dimensions that are associated with the information and (b) intrinsic dimensions that refer to the objective and native data attributes. Examples of intrinsic data quality dimensions include Accuracy, Timeliness, Consistency, and Completeness.

Data Quality Metrics: Each DQD are associated with specific metrics. A metric is an equation, or a formula developed to compute a score or the ratio of the data by quantifying its quality dimensions. The metrics provide a way to evaluate a DQD from simple formulas to more complex multivariate expressions.

Data Quality Assessment: Using a set of metrics, it is feasible to evaluate quantitatively the quality when following a data-driven strategy on existing data. For structured data, a quantitative evaluation is not possible since it is not expressed in the form of attributes with columns or rows filled with values. Therefore, unstructured data needs a different evaluation approach given the fact that we don't know how it is organized, and what are we going to assess. The introduction of a module that extracts, discover, or define attributes and features with specific DQD mapping are mandatory to proceed with quality exploration.

1.3   Data Profiling and Data Quality Profiling

Data Profiling: Data profiling can be applied at different stages of the data lifecycle. It is defined as the process of verifying data in its different types and formats such as structured, semi-structured, and unstructured data. Then collect and visualize various information about data, including: structure, patterns, statistics, metadata, data attributes or features. Moreover, all these assembled information's are used or requested for data governance, data management, and data quality control [6]. There are many data profiling analysis schemes used for this purpose, such as attribute analysis, referential analysis and functional dependency analysis [1]. In the following, we summarize the main benefits of data profiling that helps in:

1. Finding irregularities in data in the earlier stages and takes correction actions.
2. Understand content, structure, and relationships about the data in the data source.
3. Assess, validate and analyze metadata.
4. Making a statistical analysis of the data at its source.

A typical use case of data profiling is in a data cleansing process. In many commercial tools, profiling is always bundled with a data quality cleansing software. Profiling using data quality assessment tool detects data errors, such as inconsistent formatting within a column, missing values, or outliers. Therefore, profiling results might be used to measure and monitor certain quality dimensions of a dataset such as the ratios of observations that are not satisfying data constraints [2, 13].

Data Quality profiling: is the process of analyzing a dataset in the context of a quality domain defined by a set of quality and data requirements, to detect quality issues. The results of data quality profiling may include:

- Summaries describing: (a) completeness of datasets and data records (b) data problems (e.g. Wrong entries, inconsistent data) (c) problem's distribution in a dataset.
- Details about: (a) missing data records (b) data problems in existing records.

2   Literature Review

By surveying the literature, we haven't found comprehensive research that emphases Big Data quality profiling in the context of Big data. Indeed, very few works addressed the profiling process in Big Data levelling some aspects, like in [11] where the authors addressed the challenges of user profiling in Big Data using techniques focusing essentially on privacy. On the same path, the authors in [5, 10] showed that web user profiling techniques are used based on cookies for real-time profiling and marketing. They presented a framework for web user profiling based on leveraging the redundant information on the web. On the other side, Jamil and al. presented a set of guidelines for planning, conducting and reporting a systematic review and provided a review of the literature on profiling digital news for Big Data veracity [12].

In [1], Abedjan et al. stressed that data profiling is important to discover metadata, and further profiling work is required in the context of new types of data such as Big Data. Especially for profiling results visualization and interpretation that is still challenging. In [7], Naumann presented Big data profiling following the same methodology as traditional data profiling with a taste of measuring the Big Data Characteristics V's and providing some summaries. The same author in [17] revisited data profiling and stressed the need to level up to a newly modernized data profiling by developing a framework to support data profiling and motivate the need to develop new profiling techniques for Big Data.

In [6], the authors emphasized that data profiling for Big Data is very important in Data Governance. They presented various data quality metrics formulas and calculation along with the commercial and free software profiling tools used for this purpose. However, there is no detailed information on how their profiling system architecture works and process data to provide more accurate metadata. In [14, 16], the authors considered that data profiling might be used as a fast quality assessment and quality issues detection.

Most of the investigated literature hasn't addressed directly or indirectly the Big Data quality profiling. Generally, data profiling is considered as preliminary of data quality and can be considered as an introductory to data quality profiling. Also, many papers addressed the use of subsets of data to discover metadata as tradeoffs to minimize processing time and costs in the case of Big Data characteristics such as volume.

3   Big Data Quality Profiling Model

We propose a quality profiling model that inspects the quality of a dataset following different processes depending on the quality dimension to be inspected.

3.1   Big Data Quality Profiling Model Description

As illustrated in Fig. 3, The Big Data quality profiling model is architected around many modules essentially, data sampling, data profiling, the exploratory quality profiling scenarios (EQP), the quality evaluation and PPAF discovery, the data quality profile (DQP) and the QP repository.

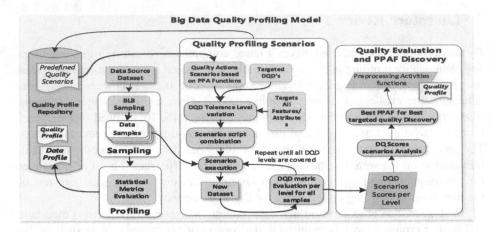

Fig. 3. Big Data Quality Profiling Model (BDQPM)

3.2   Sampling and Profiling: Statistical Metric Evaluation

We used the Bag of Little Bootstrap (BLB) [21], which combines the results of bootstrapping multiple small subsets of a Big Data dataset. The BLB algorithm uses Big Data Set to generate small samples without replacements. For each generated sample another set of samples is created by re-sampling with replacements. For data quality evaluation, we used the bootstrap since it is a re-sampling method used to gather the subsequent distribution of the whole data rather than assessing the quality of some estimators.

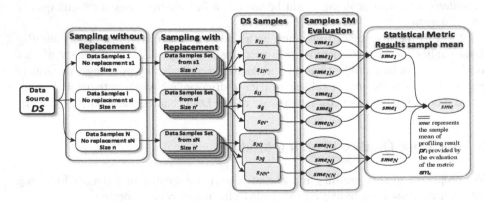

Fig. 4. Big Data sampling & statistical metric evaluation

Sampling-Profiling Description. The BDQPM has two modules handling the sampling and the profiling process. The sampling process is invoked also through BDQPM in other modules such as EQP and quality evaluation, In Fig. 4, we describe how data is sampled and profiled for a data statistic metric sm_x.

The process sequence of steps: It follows the 5 steps below:

1. Sampling of the data set *DS*, *N* bootstrap samples of **n** size without replacement s_i. *(I = 1... N)*
2. Each sample generated from step 1 is re-sampled into *N'* samples of size *n' (n' < n)* with replacements s_{ij}. *(i = 1... N, j = 1... N')*
3. For each sample s_{ij} generated in step 2, evaluate the data statistic metric sm_x as sme_{ij}.
4. For all the samples s_i, evaluate the sample mean of all *N'* samples sm_x the evaluation scores are expressed as $\overline{sme}_{ij}$
5. For the data set *DS*, evaluate the data statistic metric score $\overline{sme}$ which represents the mean of all *N* samples evaluation scores sm_x.

The Big Data Sampling-Profiling Algorithm (SPA). The sampling and profiling processes are portrayed in Algorithm 1 (SPA). The procedure *Sample_Profiling ()* is responsible for generation Big Data samples, then profiling these samples simultaneously by calculating the statistical metrics representing a certain data profile.

Algorithm 1: Big Data Sampling and Profiling(SPA)
1   **Input:** *DS* Dataset size *ss*, $A = \{a_1, .. a_k, ., a_r\}$ Attributes
2   *N* samples s_i of size *n* from *DS*
3   $SM = \{sm_0, ..., sm_c\}$ Statistical metrics set
4   **Output:** *sme : Samples Statatistical Metric Evaluation*
5   **procedure** Sample_Profiling(*DS, A, SM, N, n*)
6   //s_i a sample without and s_{ij} with replacement
7   **for each** *i* **from** *1* **to** *N*
8         $s_i \leftarrow$ Generate_Sample(*DS, N, n, no rep*)
9   **for each** *j* from *1* to *N*
10           $s_{ij} \leftarrow$ Generate Sample(*s_i, N, n, rep*)
11   **for each** *k* from *1* to *R*
12   //$sme_{ij}(k)$ *Evaluate sm_x for s_{ij} sample attribute*
13              $sme_{ij}(k) = $ Eval(*s_{ij}, sm_x, a_k, n'*)
14   **end** *k*
15   **for each** *k* from *1* to *R*
16              $sme_{ij} = sem_{ij} + sme_{ij}(k)$
17   **end** *k*
18           $sme_{ij} = sme_{ij}$ / R
19   **end** *j*
20        $sme_i = sum(sme_{ij})$ / N'
21   **end** *i*
22     $sme = sum(sme_i)/N$
23   **return** *sme*
24   **end procedure**

3.3   Quality Evaluation Module

For the quality evaluation module, a DQD is evaluated through all the data attributes using its related metric. For example, for DQD completeness, the metric will compute per attributes, per observation, the ratio of non-empty values divided by total values. It is based on the same algorithm skeleton used in Fig. 4.

3.4 Exploratory Quality Profiling for PPAF Discovery

The exploratory quality profiling component executes a series of quality scenarios based on specific DQDs, then generates quality scores. These scenarios, target attributes DQD's at each iteration based on incremental DQD acceptance ratios and applies a set of actions accordingly. In Table 1, we present some Exploratory Data Profiling scenarios, the first scenario executes the following:

Table 1. Exploratory quality profiling scenarios

DQD Tolerance Levels %			DQD	PreDefined Scenarios	Execution
min	max	step		Actions	Order
5	95	5	Completeness	DeleteCols(dqd)	1
5	95	5		DeleteRows(dqd)	1
5	95	5		DeleteCols(dqd)	1
5	95	5		DeleteRows(dqd)	2
5	95	5		DeleteRows(dqd)	1
				newdqd=Re-Evaluate()	2
				DeleteCols(newdqd)	3
5	95	5		DeleteCols(newdqd)	1
				newdqd=Re-Evaluate()	2
				DeleteRows(dqd)	3

*"**For Each** iteration of tolerance level **Repeat (Delete** Columns **with** DQD ratio less than the tolerance level **and recalculate** (the resulted dataset DQD ratio))"*.

A pre-defined scenario represents systematic quality actions applied on data attributes or observations or entities when the DQD ratios don't meet the iterative tolerance level. For each DQD required level, if the ratio is not met, the action is applied. The actions are represented as PPAF. The previous example scenario is as a script added to the QPREPO pre-defined quality scenarios table. The script might be written in R, Python, Scala, Java based on the platform used in the experimentations.

A pre-defined scenario is identified by the following information in the QPRepo:

- Scenario ID #: 001 (Key Id in the Scenarios Table)
- Target DQD ID #: 001 (Key Id in the DQD Table)
- Target DQD Name: Completeness (DQD Description)
- Target Data: [A (C), O (R), AO (CR), E] where A/C: Attribute/Column, O/R: Observation/Row, AO/CR for both, E: Entity defined as a chunk of the Dataset.
- (Scenario Script, Language) tuple for each implemented platform.

In the following, we depict the Exploratory Quality Profiling Algorithm.

```
Algorithm 2: Quality Profiling for Preprocessing Activity Function Discovery
 1 : Input: DS: Samples Set,
 2 :     QAScenarios: predefined Quality Actions Scenarios (PPA,PPF)
 3 :     DQDL: DDQD Levels ranging from min to max
 4 : Output: QResultss: Quality Results
 5 :     RRD:  Ratio of Remaining Data from the Original Data-Set DS
 6 : PPAF_Proposals_Set: Preprocessing Activity Function Set
 7 : DQDS_DS(): DQD Scores % per Data-Set
 8 : procedure EXtract_PPAF_Set(DS, QAScenarios, DQDL)
 9 :   DQDS_R(): DQD Scores % per Row, DQDS_C(): per Column
10 :   for each(DQDL_k : K from 1 to N)
11 :        procedure Compute_DQD_Scores(DS)
12 :            DQD_Scores_List<-(DQDS-R, DQDS_C, DQDS_DS)
13 :            return DQD_Scores_List
14 :        end procedure
15 :        procedure Filter_Failed_Rows_Cols(DQD_Scores_List)
16 :            for all r_i in DQDS-R
17 :                If (DQDS_R(r_i)>=DQDL_k) Add (r_i, R')
18 :            end for
19 :            for all c_j in, DQDS_C
20 :                If (DQDS_C(c_j)>=DQDL_k) Add (c_j, C')
21 :            end for
22 :            Failed_Scores_List<-(R',C')
23 :            return Failed_Scores_List
24 :        end procedure
25 :        procedure Apply_Scenarios_Scores(Failed_Scores_List, QAScenarios)
26 :            PPAF_k<-Generate_PPAF(QAScenarios, DQDL_k)
27 :            DS'<-Execute_PPAF(PPAF_k)
28 :            DQD_Scores_List'_k<-Compute_DQD_Scores(DS')
29 :        end procedure
30 :        procedure Compute_RR_Ratio(DS,DS')
31 :            RR_k =size(DS')/size(DS)* 100 %
32 :            return RR_k
33 :        end procedure
34 :        DQP<-QPRepo_Update(DQDL_k, DQDS_DS_k, DQDS_DS'_k,RR_k,PPAF_k)
35 :   end for k
36 :   for each(DQDL_k : K from 1 to N)
37 :        PPAF_Proposals_Set<-KNN Best PPAF_k per Targeted Ratios
38 :   end for k
39 :   return PPAF_Proposals_Set
40 : end procedure
```

Algorithm 2 describes the quality profiling that evaluates first the targeted DQD, then applies a set of scenarios on data that fails the DQD acceptance level. After the PPAF actions on data, another re-evaluation of DQD's is done to check new quality scores. The process is repeated until all iterations on the acceptance levels are completed (e.g. From DQD level = 5 to 95 in step 5). After all the results are gathered and listed by ratio scores, for each type of PPAF actions, a query for the best PPAF actions that leveraged the DQD scores (or lower depending on the DQD description) is performed using the KNN algorithm. The latter ranks the best PPAF combinations that can achieve the goal of the query (refer to Table 4 in the last section for more details).

4 Big Data Quality Profile Repository

The Big Data Quality Profile Repository (QPRepo) is considered as an important component of our model. All the Quality information about the data is recorded across the different modules of the framework, including the simple data profile gathered from metadata, the newly data summary, profiled data. Consequently, the QPRepo is created with the aforementioned information for structured data. The more we get into the quality profiling modules, additional information is added to the QPRepo like the quality evaluation scores of the exploratory quality profiling, the quality scores, the PPAF, the data sources, the data quality dimensions and their metrics, the pre-processing activities and their related activity functions. All this information is recorded in a repository database for Big Data quality profile management. An illustration of the QPRepo data and tables is showed in Fig. 5.

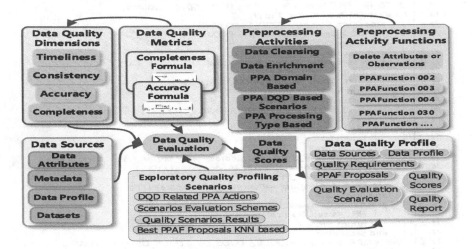

Fig. 5. Big Data quality profile repository

4.1 Quality Related Repository Components

The Data Quality Profile holds all the information about the Big Data Quality, from data and its sources, its contents (schema, metadata), its quality requirements, its quality dimensions scores, and the discovered quality pre-processing activities to enhance the quality per attributes, per DQD, or per dataset.

4.2 Data Quality Profile

As illustrated in Fig. 6, the DQP is specified in an XML document to store all the modification added through the different processes of the framework. At each module, a version of the DQP is recorded and named a data quality profile level, where the level indicates the module where the DQP was created, updated or upgraded with its pro-cesses related quality information and quality reports.

4.3 Pre-processing Activities and Related Functions (PPA+PPAF)

The PPA repository is organized as a tuple PPA (DQD, PPAF), where each data quality dimension DQD is associated with an activity function. One of the corresponding pre-processing activities is to eliminate the data that didn't satisfy this DQD or replace it with calculated values based on the data.

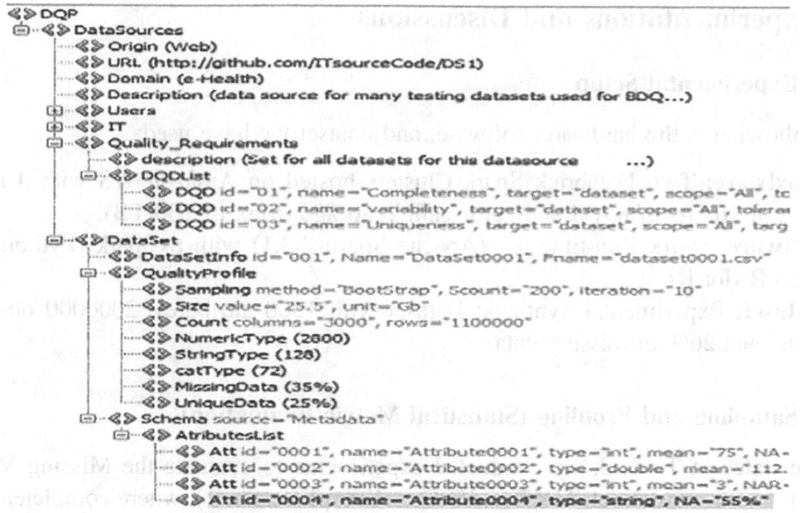

```
<> DQP
  <> DataSources
      <> Origin (Web)
      <> URL (http://github.com/ITsourceCode/DS1)
      <> Domain (e-Health)
      <> Description (data source for many testing datasets used for BDQ...)
    <> Users
    <> IT
    <> Quality_Requirements
        <> description (Set for all datasets for this datasource    ...)
      <> DQDList
          <> DQD id="01", name="Completeness", target="dataset", scope="All", tc
          <> DQD id="02", name="Variability", target="dataset", scope="All", tolera
          <> DQD id="03", name="Uniqueness", target="dataset", scope="All", targ
    <> DataSet
      <> DataSetInfo id="001", Name="DataSet0001", Fname="dataset0001.csv"
      <> QualityProfile
          <> Sampling method="BootStrap", Scount="200", iteration="10"
          <> Size value="25.5", unit="Gb"
          <> Count columns="3000", rows="1100000"
          <> NumericType (2800)
          <> StringType (128)
          <> catType (72)
          <> MissingData (35%)
          <> UniqueData (25%)
      <> Schema source="Metadata"
        <> AtributesList
            <> Att id="0001", name="Attribute0001", type="int", mean="75", NA=
            <> Att id="0002", name="Attribute0002", type="double", mean="112.
            <> Att id="0003", name="Attribute0003", type="int", mean="3", NAR=
            <> Att id="0004", name="Attribute0004", type="string", NA="55%"
```

Fig. 6. Big Data quality profile

The QPREPO is responsible for defining and managing pre-processing activities and their related functions. We illustrate QPREPO_ examples of both Preprocessing Activities (PPA) and their related PPA Functions (PPAF) in detail in Table 2 below. Each DQD is liked to a pre-processing activity which in itself has many functions that deal with the DQD issues and enhance it by removing them. For example, the DQD completeness in the table is described with the methods used to compute it, then the results formulas calculation, and its PPA category (data cleansing) with many functions (PPAF's) actions like data correction or data removal to deal with the completeness issues. The PPAF's vary from replace the missing values with several methods based on the existing data itself or drop and remove the missing data attributes, observations or a balanced combination of them as illustrated in the pre-defined scenarios in Table 2.

Table 2. Preprocessing activities and PPA functions

#	DQD	Metric	Data Type	Methods	Results*100 (%)	PPA	PPAF	PPF Related Actions or Proposals
11	Accuracy Consistency Validity	Outliers Detection	Num	Rule based	(Outliers Count)/(Total Rows), List of Outliers (Anomaly, Novelty)	Data Cleansing	Retention	Use robust classification methods
12				Linear Regression Model			Winsorizing	Replace Outliers with closest values
13				High Dimensional Outlier Detection Methods			Exlusion, Truncation	Remove Related Rows
21	Completeness	Recorded Obs	All	Count the number of Not (NA, Null or any other values that express the No Available data values)	(Not NA Count) /(Total observations (Rows))	Data Enrichment	Data Correction	Replace with Mean
22								Replace with Mode
23								Replace with Median
24							Data Removal	Remove Rows
25								Remove Columns
26								Remove Rows and Cols

5   Experimentations and Discussions

5.1   Experimental Setup

The following is the hardware, software, and dataset we have used:

- **Hardware:** Two Databrick Spark Clusters hosted on Amazon WS with 4 nodes each of 16 GB of RAM and auto-scaling storage (Max of 100 TB).
- **Software:** Spark Run-time 4.2 (Apache Spark 2.3.1) with pySpark (Python) and SparkR (for R).
- **Dataset:** Experimental Synthetic Dataset with 1300 attributes, 2000000 observations, and 26% of missing data.

5.2   Sampling and Profiling (Statistical Metric Evaluation)

As illustrated in Table 3, we conducted experiments to evaluate the Missing Values (MV) ratio for our dataset (MV% = 100% − Completeness %), where completeness is a Data Quality Dimension that can be computed for any type of tabulated attributes.

Table 3. Missing values mean using BLB sampling approach

Sample #	Num of Samples	NA Mean	Real NA	Std	p0	p25	p50	p75	p100	hist
1	25	25.91	25.65	0.56	24.77	25.56	25.89	26.25	27.14	▁▃█▃▁
2	49	25.50	25.65	0.73	23.88	25.15	25.53	25.90	27.14	▁█▇▃▁
3	100	25.71	25.65	0.86	23.73	25.17	25.60	26.42	27.36	▁▅█▅▂
4	225	25.56	25.65	0.87	23.39	24.97	25.62	26.07	27.79	▁▃█▅▁
5	400	25.68	25.65	0.78	22.97	25.20	25.73	26.18	27.67	▁▁▃█▁
6	625	25.61	25.65	0.66	23.64	25.17	25.61	26.07	27.67	▁▃█▅▁
7	900	25.55	25.65	0.79	23.32	25.04	25.55	26.07	28.32	▁▃█▂▁
8	1225	25.71	25.65	0.61	23.57	25.32	25.70	26.10	27.73	▁▂█▃▁
9	1600	25.76	25.65	0.70	23.49	25.28	25.78	26.25	28.63	▁▃█▂▁
10	2025	25.68	25.65	0.82	22.88	25.15	25.68	26.25	28.14	▁▂█▃▁

As illustrated in Fig. 7, the number of samples (sub-samples re-sampling) increases to the 10th iteration from 25 to 2025 samples with a sample mean that range between 25.50% to 25.91% of Missing Value (MV), the real missing value from the whole data is MV = 25.65% making the missing values of samples ranging between MV − 0.15 and MV + 0.41. This demonstrates the approximations based on Bootstrap sampling.

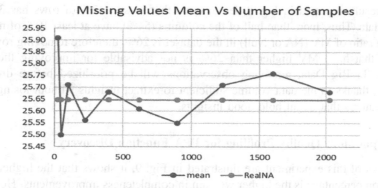

Fig. 7. Missing data representation

5.3 Missing Data Row and Column Wise

The completeness quality scores for all attributes are computed per rows and columns. An analysis of completeness of both rows and columns evaluation should highlight what is the most influential completeness acceptance level measured on rows or on columns.

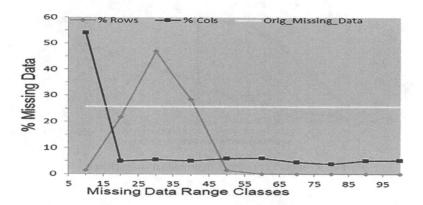

Fig. 8. Missing data ration row and column wise

Figure 8 shows the differences between the column and row-wise calculations (Attributes vs Observations). This is very important to decide which PPAF actions should be applied on data: remove first, columns or rows that have the Missing Values Threshold (MVT: Acceptance Level) less than the MVT fixed in the iterations. The possible PPAF scenarios actions that might enhance the targeted quality are: (1) Remove the objects with the percentage of missing values greater than MVT level. Either, the attributes (Columns), the instances (Observations, Rows), or a combination of both for an optimal acceptable missing value ratio and (2) Replace the missing data with the mean, median, or mode to make the qualitative assessment.

The results reported in Fig. 8 demonstrate that mostly 47% of rows has 30% of missing data. Thus, more than half of the columns (54%) have at least 10% of missing data. The ratio of MV (NA or Null) in the dataset is 26%, therefore removing rows and columns that have MV higher than 26% is not advisable for improving the completeness. In Big Data, removing observations will be prioritized before dropping attributes, thanks to the data volume sufficient to extract insights without losing some hidden features that might hold good insights.

5.4   Exploratory Quality Profiling for PPA Function Discovery

The results of this experiment are illustrated in Fig. 9, it shows that the higher MVT amputation percentage is the higher we gain in completeness improvements. However, there is a cost associated with this situation, which is the ratio of removal for both observations and features. Therefore, a reduction ratio between 100% and 70% will not be acceptable even for Big Data. For this reason, we decided that removing threshold must be higher than 50% (Redline) for the PPAF discovery and proposals as depicted in both Figs. 9, and 10.

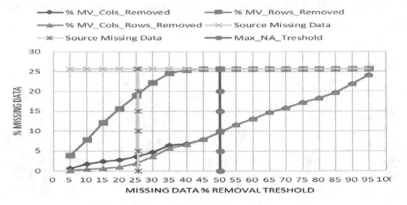

Fig. 9. Impact of removal threshold on completeness

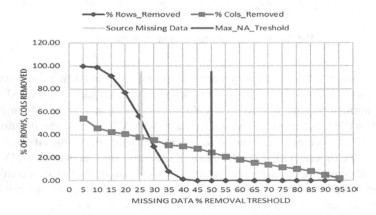

Fig. 10. Impact of removal threshold on Rows-Columns reduction

The pre-processing activity functions PPAF proposals are extracted only when a removing threshold is of 50%. Removing columns rules are more effective in completeness enforcement than rows. The removal of columns that have missing data higher than 50% achieves enhancement of completeness as the missing data drops to 10% while removing 25% of columns. The peak in Fig. 10, shows that removing rows that have more than 25% of missing values, will reduce the dataset size from 26% to 18% with 60% of rows removed, respectively. Therefore, changing the removing threshold to be higher than 30% and more will only achieve the original missing data ratio with 0% of rows reduction.

After the application of the quality scenarios and the DQD results for each iteration of acceptance levels, the next step is to select and extract the best PPAF to be used for quality improvements. The following Table 4 represents the best PPAF selection KNN Based on a set of targeted ratios parameters such as NA% ratio after columns or rows removal, and the ratios % of columns or rows Drop (might be identified as data size reduction caused by quality enhancements actions). An Example of these targeted quality ratios (in Red in Table 4) with 0% of missing values to achieve 100% of DQD Completeness in both columns and rows, an acceptance of 50% drops in rows as we are in a Big Data context letting us sacrifice observations and 10% of attributes.

The purpose is to extract the best pre-processing activity functions already selected from a general pre-defined scenario. The experimentations showed that quality profiling of big data samples is very conclusive since the best PPAF have already been enhancing the quality levels (MVT Level) with proven ratio results as illustrated in Table 4. Moreover, the extracted PPAF actions will be stored in the data quality profile (DQP) as potential functions to be applied in the pre-processing stage on the whole Big Data dataset with a certain level of confidence in the resulted data according to the targeted DQD's. The more DQD's involved the more PPAF are selected and combined to produce high-quality data for the analytics.

Table 4. KNN based best PPAF selection targeted ratios

Target %		0	0	50	10		
		NA% After					
ID	MVT Level	Cols Drop	Rows Drop	%Rows Droped	%Cols Droped	PPAF Param Rank	Euclid Dist
1	5	0.66	3.82	99.90	54.46	19	66.94
2	10	1.73	7.80	98.59	46.01	17	61.00
3	15	2.38	12.13	91.49	42.72	5	54.27
4	20	2.79	15.64	76.91	41.00	3	44.02
5	25	3.61	18.94	56.27	38.42	1	34.91
6	30	4.74	22.19	29.89	35.45	2	39.58
7	35	6.40	24.59	7.94	31.30	4	53.56
8	40	6.81	25.40	1.46	30.36	11	58.84
9	45	7.95	25.62	0.14	28.09	14	59.44
10	50	9.72	25.65	0.00	24.73	12	58.90
11	55	11.56	25.65	0.00	21.28	8	58.47
12	60	13.06	25.65	0.00	18.62	6	58.33
13	65	14.67	25.65	0.00	15.88	7	58.38
14	70	15.78	25.65	0.00	14.08	9	58.51
15	75	17.19	25.65	0.00	11.89	10	58.79
16	80	18.29	25.65	0.00	10.25	13	59.10
17	85	19.68	25.65	0.00	8.29	15	59.57
18	90	21.92	25.65	0.00	5.16	16	60.51
19	95	24.12	25.65	0.00	2.11	18	61.66

6 Conclusion

Big Data has emerged as a paradigm for extracting insights from huge amounts of data. However, data quality is considered a key for its acceptance and adoption as a poor data quality might severely affect the Big data analysis results. In this paper, we identified the key research challenges in evaluating Big data quality. We proposed a Big data profiling model to cope with data quality in an early stage of the Big Data lifecycle by providing a set of actions to be implemented in the pre-processing phase to ensure a high-quality related dataset. The set of experimentations we have conducted have validated key features of our BDQPM model including sampling and profiling, sampling and quality evaluation, and exploratory quality profiling. The results we have obtained demonstrated that quality profiling strengthens the efficiency of the pre-processing and processing phases of Big Data. Finally, we are planning to extend our BDQPM to cover and discover quality profiling rules for unstructured data, that represent almost 80% of the overall Big data, where businesses and companies are highly interested to explore and get valuable insights.

References

1. Abedjan, Z.: An introduction to data profiling. In: Zimányi, E. (ed.) eBISS 2017. LNBIP, vol. 324, pp. 1–20. Springer, Cham (2018). https://doi.org/10.1007/978-3-319-96655-7_1
2. Abedjan, Z.: Data profiling. In: Sakr, S., Zomaya, A. (eds.) Encyclopedia of Big Data Technologies, pp. 563–568. Springer, Cham (2018). https://doi.org/10.1007/978-3-319-77525-8_8
3. Assunção, M.D., Calheiros, R.N., Bianchi, S., Netto, M.A.S., Buyya, R.: Big data computing and clouds: Trends and future directions. J. Parallel Distrib. Comput. **79**(C), 3–15 (2015). https://doi.org/10.1016/j.jpdc.2014.08.003
4. Batini, C., Cappiello, C., Francalanci, C., Maurino, A.: Methodologies for data quality assessment and improvement. ACM Comput. Surv. **41**, 1–52 (2009)
5. Chester, J.: Cookie wars: how new data profiling and targeting techniques threaten citizens and consumers in the "Big Data" era. In: Gutwirth, S., Leenes, R., De Hert, P., Poullet, Y. (eds.) European Data Protection: in Good Health, pp. 53–77. Springer, Dordrecht (2012). https://doi.org/10.1007/978-94-007-2903-2_4
6. Dai, W., Wardlaw, I., Cui, Yu., Mehdi, K., Li, Y., Long, J.: Data profiling technology of data governance regarding big data: review and rethinking. Information Technology: New Generations. AISC, vol. 448, pp. 439–450. Springer, Cham (2016). https://doi.org/10.1007/978-3-319-32467-8_39
7. Naumann, F.: Big Data Profiling (2014)
8. Géczy, P.: Big data characteristics. The Macrotheme Review **3**, 94–104 (2014)
9. Glowalla, P., Balazy, P., Basten, D., Sunyaev, A.: Process-driven data quality management – an application of the combined conceptual life cycle model. Presented at the 2014 47th Hawaii International Conference on System Sciences (HICSS), pp. 4700–4709 (2014). https://doi.org/10.1109/HICSS.2014.575
10. Gu, X., et al.: Profiling Web users using big data. Soc. Netw. Anal. Min. **8**, 24 (2018). https://doi.org/10.1007/s13278-018-0495-0

11. Hasan, O., Habegger, B., Brunie, L., Bennani, N., Damiani, E.: A discussion of privacy challenges in user profiling with big data techniques: the EEXCESS use case. In: BigDataCongress, pp. 25–30 (2013)
12. Eembi, N.B.C., Ishak, I.B., Sidi, F., Affendey, L.S., Mamat, A.: A systematic review on the profiling of digital news portal for big data veracity. Proc. Comput. Sci. **72**, 390–397 (2015)
13. Johnson, T.: Data profiling. In: Liu, L., Özsu, M.T. (eds.) Encyclopedia of Database Systems, pp. 808–812. Springer, New York (2018). https://doi.org/10.1007/978-1-4614-8265-9
14. Loshin, D.: Rapid Data Quality Assessment Using Data Profiling, vol. 15 (2010)
15. Maier, M., Serebrenik, A., Vanderfeesten, I.T.P.: Towards a Big Data Reference Architecture. University of Eindhoven (2013)
16. McNeil, B.J., Pedersen, S.H., Gatsonis, C.: Current issues in profiling quality of care. Inquiry **29**, 298–307 (1992)
17. Naumann, F.: Data profiling revisited. ACM SIGMOD Rec. **42**, 40–49 (2014)
18. Oliveira, P., Rodrigues, F., Henriques, P.R.: A formal definition of data quality problems. In: IQ (2005)
19. Prabha, M.S., Sarojini, B.: Survey on Big Data and Cloud Computing, pp. 119–122. IEEE (2017)
20. Sidi, F., Shariat Panahy, P.H., Affendey, L.S., Jabar, M.A., Ibrahim, H., Mustapha, A.: Data quality: a survey of data quality dimensions. In: CAMP 2012, pp 300–304 (2012)
21. Talwalkar AKA The Big Data Bootstrap. 20
22. Sun, Z.: 10 Bigs: Big Data and Its Ten Big Characteristics (2018). https://doi.org/10.13140/rg.2.2.31449.62566

On Development of Data Science and Machine Learning Applications in Databricks

Wenhao Ruan⑩, Yifan Chen⑩, and Babak Forouraghi⁽⊠⁾⑩

Saint Joseph's University, Philadelphia, PA 19131, USA
{wr689995,yc689980,bforoura}@sju.edu

Abstract. Databricks is a unified analytics engine that allows rapid development of data science applications using machine learning techniques such as classification, linear and nonlinear regression, clustering, etc. Existence of myriad sophisticated computational options, however, can become overwhelming for designers as it may not always be clear what choices can produce the best predictive model given a specific data set. Further, the mere high dimensionality of big data sets is a challenge for data scientists to gain a deep understanding of the results obtained by a utilized model.

This paper provides general guidelines for utilizing a variety of machine learning algorithms on the cloud computing platform, Databricks. Visualization is an important means for users to understand the significance of the underlying data. Therefore, it is also demonstrated how graphical tools such as Tableau can be used to efficiently examine results of classification or clustering. The dimensionality reduction techniques such as Principal Component Analysis (PCA), which help reduce the number of features in a learning experiment, are also discussed.

To demonstrate the utility of Databricks tools, two big data sets are used for performing clustering and classification. A variety of machine learning algorithms are applied to both data sets, and it is shown how to obtain the most accurate learning models employing appropriate evaluation methods.

Keywords: Big data · Machine learning · Cloud computing · Classification · Data Science · Clustering · Databricks

1 Introduction

Along with the explosion of big data on the Internet, the interest in the field of Data Science is rapidly increasing. As effective tools, Data Science utilizes artificial intelligence (AI) and machine learning (ML) to efficiently and accurately discover meaningful patterns hidden in large volumes of data. In other words, Data Science uses AI and ML to figure out best solutions to real-world problems in novel ways [3].

ML, as a subfield of AI, teaches a machine how to learn. More specifically, with its use of methodologies from many various disciplines such as statistics, operations research, mathematics, etc., it automatically builds analytical models to discover hidden patterns and relations in data without being explicitly programmed [24].

© Springer Nature Switzerland AG 2019
Y. Xia and L.-J. Zhang (Eds.): SERVICES 2019, LNCS 11517, pp. 78–91, 2019.
https://doi.org/10.1007/978-3-030-23381-5_6

Nowadays, ML and statistical analysis are converging more and more. Both of these methodologies heavily utilize pattern recognition and data mining. In fact, over the past decades, these two data-driven disciplines have complemented each other and will most likely continue the same trend for years to come [1].

In addition to predictive analytics, data visualization tools are important to help data scientists better understand the significance of data. The underlying patterns and correlations buried deep underneath numbers and words can be easily revealed by the power of data visualization software like Tableau, Qlik and D3.js [12].

Cloud computing is a form of service-oriented computation, and it has allowed an ever increasing number of applications, services and platforms to be available to the general public [5]. ML benefits well from cloud computing because of the advantages of low cost of operations, scalability, and the necessary processing power to analyze large volumes of data. Databricks is an example of a just-in-time cloud-based platform. It was created to help users with carrying out the processes from data preparation to experimentation, and also with quick deployment of ML application. Databricks is a hundred times faster than the open source Apache Spark [13].

Databricks allows access to a rich set of ML algorithms ranging from clustering to SVM and various forms of linear/non-linear regression. As an example, K-means is a common unsupervised learning algorithm that has the tendency to get stuck in local optima. But, hierarchical clustering is often described as a better clustering approach in quality. Bisecting K-Means is a combination of K-Means and hierarchical clustering, as it combines the run-time efficiency of "regular" K-means with the higher quality of hierarchical clustering [6]. In ML, a kernelized SVM (Support Vector Machine), can perform inefficiently when processing large data sets (over 100K). The main cause is that kernelized SVM requires the computation of a distance function between each point in the dataset, which could require $O(n_{features} * n_{observations}^2)$ operations. To improve that, techniques such as data normalization, Stochastic Gradient Descent (SGD), as well as kernel approximation are often applied [2].

One of the most challenging problems for novice data scientists is to determine which algorithms are best suited for which data sets. Many aspects of the task such as the size of the data, the needed accuracy of the results, and the available computational time and resources must be well taken into consideration. The main contribution of this paper is to offer guidelines and concrete case studies to data scientists who are interested in working with Databricks. Specifically, the SAS Algorithm Flowchart is discussed which provides useful tips for solving specific problems. The main goal is to allow the user to easily find the appropriate algorithm depending on the speed, accuracy, and to evaluate the significance of the obtained results [10]. Further, the developed guidelines for using the scikit suite are included. These guidelines offer a clear view on how to select an algorithm based on the size of the dataset.

The remainder of the paper is organized as follows. Section 2 is a brief introduction to ML model building. Section 3 discusses the conducted machine learning experiments conducted on two big data sets in Databricks. Specifically, the experiments show how to deal with a dataset that has no labels by using an unsupervised learning algorithm. They also demonstrate the performance of a standard (unoptimized) kernelized SVM, comparing it with other optimization techniques. Finally, Sect. 4 is the conclusions and future directions.

2 Building ML Models

Figure 1 depicts a simple Data Science workflow from the original dataset to analyzing the obtained classification accuracy. First, Data Cleaning is performed, which is an essential process before training. During this phase, necessary actions take place to deal with missing or corrupt values or to detect outliers in order to improve overall model accuracy. Second, the dataset is split for model training and accuracy testing. Depending on the specific dataset, operations such as StringIndexer, VectorAssembler, and OneHotEncoder can be applied to dataset. StringIndexer encodes a string column to a column of label indices. VectorAssembler is a transformer that combines a given list of columns into a single vector column. OneHotEncoder maps a column of label indices to a column of binary vectors, allowing algorithms which expect continuous features, such as Logistic Regression, to use categorical features [9]. The next step is to feed the ML model with the training set, and compute the confusion matrix, which reports the obtained classification accuracy. If the accuracy doesn't meet the requirements of the problem and the domain, try adjusting the corresponding parameters of ML model.

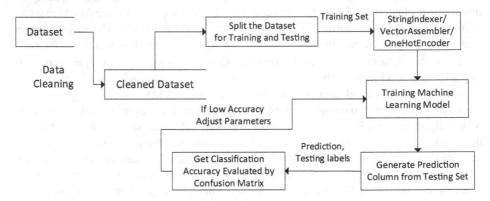

Fig. 1. Workflow of conducting a ML model training

For Machine Learning beginners, many times it is difficult to find the right estimator to solve a specific classification problem. When dealing with different types of data, different estimators have their own strengths and weaknesses. The hierarchical graph below is designed to give them a bit of a rough guide on which estimators to use to perform classification.

Similarly, Fig. 3 shows which algorithms work best when performing regression. Both Figs. 2 and 3 are based on the scikit-learn algorithm cheat-sheet, which includes various ML techniques such as regression, clustering, and dimensionality reduction [4, 20].

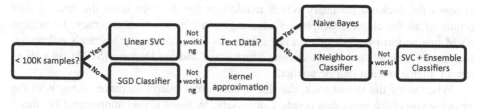

Fig. 2. Classification map guide

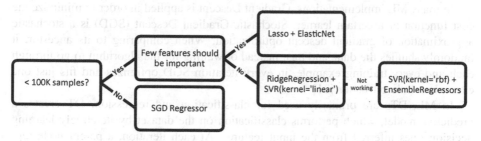

Fig. 3. Regression map guide

3 The Experiments

The experiments discussed in this section demonstrate the typical process of training a big data ML model and evaluating the obtained accuracy in Databricks. Specifically, the experiments were designed using two very large datasets and three machine learning algorithms. The following sections provide the details of the utilized algorithms, the necessary configurations, and the analysis of the obtained results.

3.1 Machine Learning Algorithms

Bisecting K-means is an unsupervised learning algorithm, which is a combination of the traditional K-Means and hierarchical clustering. Each cluster split results in the lowest aggregated Sum of Squared Errors (SSE). The key point is that Bisecting K-means can converge to a global optimum, instead of getting stuck in local optimum as it happens in K-means. Additionally, Bisecting K-means is much more efficient than basic K-means [14, 19].

Back Propagation Neural Networks (BPNN) can learn complex nonlinear function mappings even with a large number of features. Decision Trees (DT) can provide a clear explanation of the learned decision process (the tree). However, Neural Networks cannot provide any explanation for the extracted classification or regression model. They are in a sense "black boxes" because of lack of transparency in the nature of the obtained results [16].

A Support Vector Machine (SVM) is a supervised machine learning model, which discriminates different categories by a separating line, or a hyperplane, in non-linear classification problems. It differs from other classification algorithms in the way that it chooses the decision boundary which maximizes the distance from the nearest data points of all the classes. Despite its advantages in classification accuracy, kernelized SVM can perform inefficiently on large data sets. To improve the learner's efficiency, techniques such as data normalization, Stochastic Gradient Descent (SGD), as well as kernel approximation can be applied.

When using the kernel trick, the size of the kernel matrix increases along with the growing size of the input data points. Conversely, by using kernel approximation, those data points are instead projected onto some approximated lower dimensional space, which saves a lot of time since the matrix is now much smaller [2].

In many ML implementations, Gradient Descent is applied in order to minimize the cost function of a certain learner. Stochastic Gradient Descent (SGD) is a stochastic approximation of gradient descent optimization. When comparing to its ancestor, it randomly shuffles the data, and then instead of waiting for the algorithm to go through each and every training example, in every iteration SGD optimizes and fits just one example a little bit better [21].

In ML, DTs are primarily used for classification and regression. DT creates a predictive model, which performs classification on the dataset by iteratively learning decision rules inferred from the input features. At each iteration, a parent node represents a test on some input feature, and its leaves represent the classification result. DT performs well with large data sets, as the cost is logarithmic in the number of data points used to train the tree [7].

Scikit-learn offers various classification metrics for model evaluation. In this work, both the confusion matrix and classification report are used to evaluate the actual versus the predicted outcomes. The classification report is a built-in tool, which includes main classification metrics, consisting of precision, recall, f_1-score and support values (number of predicted instances).

Both sets of experiments were conducted on Databricks clusters configured as shown below (Table 1):

Table 1. Cluster configurations

Driver instances memory/Core/DBU	6 GB memory, 0.88 Cores, 1 DBU
Databricks runtime version	Runtime:5.0 (Scala 2.11, Spark 2.4.0)
Python version	2

3.2 Experiment 1

The dataset used in the first experiment relates to individual household electric power consumption, including over two million measurements gathered in a house located in Sceaux (near Paris, France) between December 2006 and November 2010 (47 months), with a one-minute sampling rate [11]. In this dataset, the date and time columns are removed, since they had no relevance on the performance of the learning model, thus, the remaining 7 features are shown in Table 2.

The Bisecting K-means algorithm requires that the dataset has no missing values. Thus, data cleaning was performed to remove all instances containing nulls. When

Table 2. Individual household power consumption features

Variable name	Description
global_active_power	Household global minute-averaged active power (in kilowatt)
global_reactive_power	Household global minute-averaged reactive power (in kilowatt)
voltage	Minute-averaged voltage (in volt)
global_intensity	Household global minute-averaged current intensity (in ampere)
sub_metering_1	Energy sub-metering No. 1 (in watt-hour of active energy)
sub_metering_2	Energy sub-metering No. 2 (in watt-hour of active energy)
sub_metering_3	Energy sub-metering No. 3 (in watt-hour of active energy)

deploying Bisecting K-means algorithm, there is one primary parameter needed to be adjusted, K, which is the desired number of clusters. The obtained clustering results generated by different values of K are shown in Table 3.

Table 3. Sum of squared errors of different values of K

K	SSE	Time consumed (minutes)
2	1.96E8	1.85
3	1.71E8	2.02
4	1.16E8	2.03
5	1.09E8	2.94
6	1.04E8	3.16
7	6.36E7	3.13
8	4.32E7	3.68
9	4.12E7	4.39
10	3.80E7	4.15
11	3.62E7	4.39

The elbow method is the oldest visual method for determining the appropriate number of clusters in a data set. But, sometimes it results in a curve that is continuously descending, and that make it much more ambiguous to find the elbow point [22]. Figure 4 is the line chart of SSE of different values of K. Since the spot where K = 7, looks mostly like an elbow, while 7 is selected as the number of clusters.

After clustering, classification is performed on the resulting dataset (7 classification labels) using the modified Kernel SVM (approximated with RBFSampler and fed into SGDClassifier). The SGDClassifier built in with the scikit-learn library, contains linear classifiers – SVM, logistic regression and a.o., with SGD training [21].

In detail, the training data (80% of total) was scaled to the range [− 1, 1], in order to improve SVM's performance. RBFSampler was utilized to approximate the feature map of an RBF kernel, since the dataset was too massive for traditional Kernel SVM learning. Finally, the prepared data was fed to a linear SVM (with SGD) and learning was performed [23]. The obtained classification accuracy evaluated by a confusion matrix on a testing set of size 200,000 is shown in Table 4.

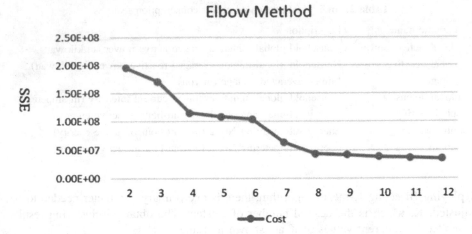

Fig. 4. Elbow method

Table 4. Confusion matrix

Cluster no.	Precision	Recall	f_1-score	Support
0	0.99	0.99	0.99	67,553
1	0.99	0.99	0.99	62,713
2	0.92	0.94	0.93	4,062
3	0.98	0.99	0.98	37,745
4	0.99	0.96	0.97	20,455
5	0.97	0.95	0.96	4,071
6	0.98	0.96	0.97	3,401
avg/total	**0.99**	**0.99**	**0.99**	**200,000**

In comparison to standard Kernel SVM, the enhanced method took only 15.68 s in average for training and testing, whereas the former needed more than 3 h of training. All 5 runs along with the time consumed are recorded in Table 5.

Table 5. Enhanced SVM run-times

Run	Time (in seconds)
1	15.29
2	15.40
3	16.68
4	15.52
5	15.49
avg/total	**15.68**

For visualization purposes, dimensionality reduction (from 7 features to 2) is essential, therefore a 2-dimensional graph was drawn by using PCA (Principal Component Analysis) [18]. As shown in Fig. 5, the resulting classification plot looks somewhat different from the original one; this is due to the fact that only 60% of the original information is retained, after the process of PCA [8].

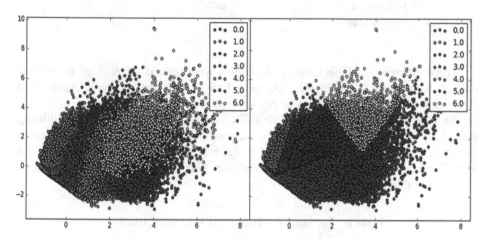

Fig. 5. Scatter plot before (left) and after (right) classification

3.3   Experiment 2

The dataset used in the second experiment is Physical Activity Monitoring dataset, which contains 3.8 million rows of data collected from 9 subjects wearing 3 inertial measurement units and a heart rate monitor [17]. The dataset's null records were removed, and a total of 53 features were selected along with the classification outcome ActivityID as shown in Table 6.

Table 6. Physical activity monitoring features

Index	Description	Type
1	Timestamp	Continuous (in seconds)
2	ActivityID	25 classification labels
3	Heart Rate	Continuous (in bpm)
4–20	IMU hand	See Table 7
21–37	IMU chest	See Table 7
38–54	IMU ankle	See Table 7

The unnecessary columns that had no relevance on the performance of the model (columns 14–17 in IMU sensory data) were removed, and the rest of the columns are shown in Table 7.

Table 7. IMU sensory data

Index	Description	Type
1	Temperature	Continuous (in °C)
2–4	3D-acceleration data, scale, resolution	Continuous (in ms^{-2}), ± 16 g, 13-bit
5–7	3D-acceleration data, scale, resolution	Continuous (in ms^{-2}), ± 16 g, 13-bit
8–10	3D-gyroscope	rad/s
11–13	3D-magnetometer	μT

Three different machine learning algorithms were trained on 80% of the overall dataset: SVM with SGDClassifier, Decision Tree and Neural Network. Again, data cleaning was conducted by dropping rows which contained null cells.

As done in Experiment 1, SVM with SGDClassifier was applied instead of utilizing Kernel SVM, and the confusion matrix was computed on a 20% testing dataset is shown in Table 8. Because of the limited space, only 8 of 25 classifications are shown here. The average training of enhanced SVM over 5 statistically independent runs took 6.60 min (Table 9).

Table 8. Confusion matrix of SVM with SGDClassifier

Cluster no.	Precision	Recall	f_1-score	Support
0	0.53	0.61	0.57	20,465
1	0.95	0.94	0.95	3,582
2	0.79	0.73	0.76	3,400
3	0.56	0.69	0.62	3,474
4	0.53	0.73	0.62	4,108
5	0.60	0.68	0.64	1,712
...				
24	0.81	0.02	0.05	874
avg/total	**0.63**	**0.63**	**0.60**	**69,758**

Table 9. SVM with SGDClassifier run-times

Run	Time (in seconds)
1	7.90
2	6.39
3	6.39
4	6.15
5	6.16
avg/total	**6.60**

Table 10. Confusion matrix of MLP

Cluster no.	precision	recall	f_1-score	Support
0	0.84	0.87	0.85	20,566
1	0.98	0.95	0.97	3,563
2	0.95	0.96	0.95	3,365
3	0.92	0.95	0.93	3,485
4	0.91	0.93	0.92	4,159
5	0.90	0.84	0.87	1,711
6	0.94	0.94	0.94	2,990
...				
24	0.83	0.72	0.77	878
avg/total	**0.89**	**0.89**	**0.89**	**69,758**

Since the obtained overall accuracy was only 63%, K-fold validation was used but that resulted in 41% accuracy. The two other ML algorithms were applied to the same dataset as following.

Multi-layer Perceptron (MLP) is a built-in Neural Network model in scikit-learn library, which trains using back propagation. The model optimizes the log-loss function using LBFGS or stochastic gradient descent (SGD) [15]. Before training the model, MinMaxScaler was used to increase SVM speed and eliminate outlier data in the original dataset. The cleaned data was then fed into the MLP with two hidden layers, one has 30 neurons and another has 6 neurons. The obtained classification accuracy evaluated by confusion matrix used 20% of the testing dataset as shown in Table 10. The average training of MLP over 5 statistically independent runs took 5.40 min (Table 11).

Table 11. MLP run-times

Run	Time (in minutes)
1	4.55
2	5.57
3	5.01
4	5.88
5	5.98
avg/total	**5.40**

Scikit-learn library also has a model for Decision Tree, called DecisionTreeClassifier which is capable of both performing binary and multi-class classification. DecisionTreeClassifier takes two arrays as input – the features and the class labels for training samples. Before fitting the training samples, the parameters "max_depth = 20, min_samples_split = 2, criterion = 'entropy'" were set for high classification accuracy. The average training of DT over 5 statistically independent runs took 1.89 s, which was the fastest among all three classification algorithms, and the resulting confusion matrix is shown in Table 12 (Table 13).

Table 12. Confusion matrix of decision tree

Cluster No.	Precision	Recall	f_1-score	Support
0	0.97	0.98	0.97	20,567
1	1.00	1.00	1.00	3,504
2	0.99	0.99	0.99	3,303
3	0.99	1.00	0.99	3,406
4	0.97	0.97	0.97	4,248
5	0.99	0.98	0.98	1,700
6	0.99	0.98	0.98	2,958
...				
24	0.98	0.94	0.96	883
avg/total	**0.98**	**0.98**	**0.98**	**69,758**

Table 13. Decision tree run-times

Run	Time (in seconds)
1	1.72
2	1.64
3	2.03
4	2.04
5	2.03
avg/total	**1.89**

To compare, the Neural Network model was the slowest because the back propagation requires lengthy computations. Decision Tree performed well on the highly nonlinear data, and it was the fastest to build and test. Table 14 summarizes the accuracy measures and time requirements for the three classifiers.

Table 14. Positive predictive value (PPV)

Learning model	Accuracy	Time consumed
Kernel SVM with SGD	63%	6.60 s
Neural Network hidden layers – 30, 6	89%	5.40 min
Decision Tree Max Depth = 20	98%	1.89 s

As shown in Table 14, the best accuracy was obtained by the decision tree learner while Kernel SVM with SGD produced the least favorable result. In term of future investigations, another study will have to be conducted to determine the exact cause of performance variation among different ML techniques.

And finally, to provide a visualization aid to designers, PCA was conducted for dimensionality reduction from 41 features to 2. This resulted in 38% loss of original information, and the results are shown in Fig. 6.

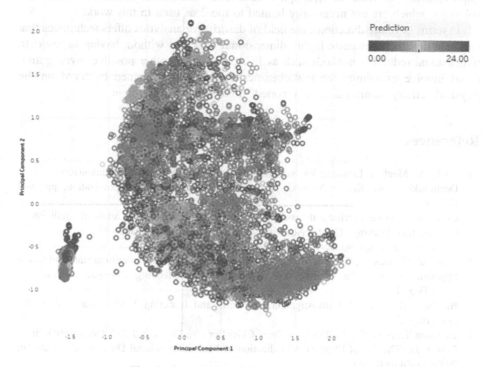

Fig. 6. Scatter plot of decision tree prediction

4 Conclusions

This paper provided general guidelines for utilizing a variety of machine learning algorithms on the cloud computing platform, Databricks. Visualization is an important means for users to understand the significance of the underlying data. Therefore, it was also demonstrated how graphical tools such as Tableau can be used to efficiently examine results of classification or clustering. The dimensionality reduction techniques such as Principal Component Analysis (PCA), which help reduce the number of features in a learning experiment, were also discussed.

To demonstrate the utility of Databricks tools, two big data sets were used for performing clustering and classification. The first experiment compared regular kernel SVM with optimized kernel SVM, where there was a significant difference in classification performance. To highlight differences in obtained accuracies and time requirements of various algorithms, another very large dataset and three different supervised classification algorithms were used. The obtained results confirmed that the Decision Tree algorithm was significantly more accurate on highly non-linear data.

In order to better visualize highly dimensional data sets, the PCA technique was utilized to reduce the number of dimensions from 7 to 2. However, this dimension reduction resulted in 40% loss of accuracy in overall classification. In terms of future improvements, the field of Descriptive Analytics offers other dimension reductions solutions, which are not necessarily limited to the 2 as used in this work.

In terms of future directions, the field of descriptive analytics offers techniques that can help designers visualize highly dimensional data sets without having to resort to dimensional reduction methods such as PCA. Further, another possible investigation could involve examining the low classification accuracy obtained by SVM on the physical activity monitor data set reported in the second experiment.

References

1. Shah, A.: Machine Learning Vs. Statistics. Accessed 21 Dec 2018. Documentation
2. Deshmukh, A.A.: Kernel Approximation, Stats 608 Methods in Optimization, pp. 1–3 (2015)
3. Custer, C.: Answer to what is the connection between data science and artificial intelligence? Is it machine learning? (2016). Accessed 21 Dec 2018. Documentation
4. Choosing the right estimator: - scikit-learn 0.20.1 Documentation
5. Agrawal, D., Das, S., EI Abbadi, A.: Big data and cloud computing: current state and future opportunities. In: 14th International Conference on Extending Database Technology, pp. 530–533 (2011)
6. Blodgett, D.: An initial investigation: K-Means and bisecting K-Means algorithms for clustering (2016)
7. Decision Trees (DTs). In: Decision Trees Classifier - scikit-learn 0.20.1 Documentation
8. Sarkar, D.: The Art of Effective Visualization of Multi-dimensional Data. Accessed 20 Jan 2019. Documentation
9. Extracting, transforming and selecting features. - scikit-learn 0.20.1 Documentation
10. Li, H.: Which machine learning algorithm should I use. The SAS Data Science Blog (2017). Documentation
11. Individual household electric power consumption Data Set. In: UCI Machine Learning Repository. Documentation
12. Rouse, M.: Data Visualization. Accessed 21 Dec 2018. Documentation
13. Rouse, M.: What is DataBricks? (2018). Documentation
14. Steinbach, M., Karypis, G., Kumar, V.: A Comparison of Document Clustering Techniques
15. Multi-layer Perceptron (MLP). In: Multi-layer Perceptron - scikit-learn 0.20.1 Documentation
16. Donges, N.: Pros and Cons of Neural Networks (2018). Documentation
17. PAMAP2 Physical Activity Monitoring Data Set. In: UCI Machine Learning Repository. Accessed 16 Oct 2018. Documentation
18. Principal component analysis (PCA). - scikit-learn 0.20.1 Documentation
19. Ruchika R. Patil, Amreen Khan: Bisecting K-means for Clustering Web Log data (2015), International Journal of Computer Applications (0975 – 8887), Volume 116 – No. 19
20. Asthana, S.: You need these cheat sheets if you're tackling Machine Learning Algorithms (2017)
21. Stochastic Gradient Descent (SGD). In: SGDClassifier - scikit-learn 0.20.1 Documentation

22. Kodinariya, T.M., Makeana, P.R.: Int. J. Adv. Res. Comput. Sci. Manag. Stud. (2017). ISSN 2347-1778, 2321-7782
23. Malik, U.: Implementing SVM and Kernel SVM with Python's Scikit-Learn (2018)
24. Thompson, W., Li, H., Bolen, A.: Artificial intelligence, machine learning, deep learning and beyond – understanding AI technologies and how they lead to smart applications. Accessed 21 Dec 2018. Documentation

Concept Drift Adaptive Physical Event Detection for Social Media Streams

Abhijit Suprem[1(✉)], Aibek Musaev[2], and Calton Pu[1]

[1] Georgia Institute of Technology, Atlanta, GA 30332, USA
asuprem@gatech.edu
[2] University of Alabama, Tuscaloosa, AL 35487, USA

Abstract. Event detection has long been the domain of physical sensors operating in a static dataset assumption. The prevalence of social media and web access has led to the emergence of *social*, or *human* sensors who report on events globally. This warrants development of event detectors that can take advantage of the truly dense and high spatial and temporal resolution data provided by more than 3 billion social users. The phenomenon of *concept drift*, which causes terms and signals associated with a topic to change over time, renders static machine learning ineffective. Towards this end, we present an application for physical event detection on social sensors that improves traditional physical event detection with concept drift adaptation. Our approach continuously updates its machine learning classifiers automatically, without the need for human intervention. It integrates data from heterogeneous sources and is designed to handle weak-signal events (landslides, wildfires) with around ten posts per event in addition to large-signal events (hurricanes, earthquakes) with hundreds of thousands of posts per event. We demonstrate a landslide detector on our application that detects almost 350% more landslides compared to static approaches. Our application has high performance: using classifiers trained in 2014, achieving event detection accuracy of 0.988, compared to 0.762 for static approaches.

Keywords: Concept drift · Machine learning event detection ·
Disaster detection

1 Introduction

The ubiquitous presence of web data and increase in users sharing information in social media has created a global network of *human reporters* who report on live events. Such human reporters can be considered as social sensors that provide information about live physical events around the globe [4–7]. Development of applications that can take advantage of social sensors to perform physical event detection are a clear next step. The primary challenge lies in the actual event detection: since social sensor data is a noisy text stream, machine learning models are required. Further, there is the phenomena of concept drift, where the distribution of real-world data changes with respect to time. This is especially apparent with text data. We provide an example with *landslide* detection on social media: the word *landslide* can refer to the disaster event or

© Springer Nature Switzerland AG 2019
Y. Xia and L.-J. Zhang (Eds.): SERVICES 2019, LNCS 11517, pp. 92–105, 2019.
https://doi.org/10.1007/978-3-030-23381-5_7

elections, among others. Since October and November are often election seasons in the United States, classification models that are tuned to ignore social media data with election related landslide keywords are more appropriate in this data window. In other months, the presence of election-related landslide tweets is scarce, which can cause increase false negatives in models overfit for election-related landslide tweets. This instance of changing data distributions is a form of concept drift.

So, social sensors constitute a challenge for the traditional approaches to text classification, which involve static machine learning classifiers that are never updated. Event detection systems for social sensors without concept drift adaptation face performance deterioration. We can consider Google Flu Trends (GFT) as an example; GFT was originally created to identify seasonal trends in the flu season [9]. However, the models did not incorporate changes in Google's own search data, causing increasing errors after release [9–11]. Our application addresses these challenges by incorporating concept drift adaptation. Additionally, we develop *automated* concept drift adaptation techniques to automatically generate training data for machine learning model updates. This is necessary due to the sheer volume of social media data – it is impractical to manually label the millions of social media posts per day. Our approach allows us to perform drift adaptation without human labelers, which significantly reduces training bottlenecks.

Specifically, we have the following contributions:

1. We present a drift-adaptive event detection application that performs physical event detection on social sensors using machine learning. We also show automated classifier updates for concept drift adaptation.
2. We develop a procedure to combine news articles and physical sensor data (e.g. rainfall data from NOAA and earthquake data from USGS) to perform automated training data generation for concept drift adaptation. Our application uses the low-latency, abundant social sensor data to perform physical event detection with machine learning classifiers, and the high-latency, scarce physical sensor data to tune and update classifiers.

We demonstrate our event detection application with landslide detection. We select landslides because they do not have dedicated physical sensors (in contrast to tsunamis or earthquakes); however, they cause large monetary and human losses each year. Landslides are a also what we call a *weak-signal* disaster: landslide-related social media data has significant noise in social media streams. Also, usage of *landslide* keywords to reference disasters is small compared to usage to reference irrelevant topics such as election landslides.

We compare our landslide detection application (LITMUS-adaptive) to the static approach in [2], which we call LITMUS-static. Our approach, LITMUS-adaptive, detects 350% more landslide events than LITMUS-static. We also evaluate our adaptive classifiers' accuracy compared to the static classifiers in LITMUS-static. We train classifiers with data in 2014, and compare performance of static and adaptive approaches in 2018. LITMUS-adaptive has f-score of 0.988 in 2018, compared to f-score of 0.762 in LITMUS-static, showcasing improvements our drift adaptive approach makes.

The rest of the paper is organized as follows: Sect. 2 covers related work. Section 3 covers data sources used in our application. Sections 4 and 5 provide implementation details for our application. Sections 6 and 7 evaluate our application quantitatively and qualitatively, respectively. Section 8 presents our conclusions.

2 Related Work

2.1 Physical Event Detection on Social Sensors

Earthquake detection using social sensors was initially proposed in [1]. There have also been attempts to develop physical event detectors for other types of disasters, including flooding [2], flu [3, 4], infectious diseases [5], and landslides [6, 7]. In most cases, the works focus on large-scale disasters or health crises, such as earthquakes, hurricanes [8], and influenza that can be easily verified and have abundant reputable data Our application is general purpose, as it can handle small-scale disaster such as landslides and large-scale disasters. The existing approaches also assume data without concept drift. However such assumptions, made in Google Flu Trends [9, 10] degrade in the long term.

2.2 Concept Drift Adaptation

Recent drift adaptation approaches evaluate their methods with synthetic data [11–14]. Such data is perturbed to include specific, known forms of drift. Several mechanisms have been developed for handling concept drift with numeric, sensor data.

Windowing, or sliding windows, is a common technique for adaptation. This approach uses multiple data memories, or windows of different lengths sliding over incoming data. Each window has an associated model. The SAM-KNN approach uses k-NN classifier to select window closest to a new data sample for classification [15]. Nested windows are considered in [16] to obtain training sets.

Adaptive Random Forests augment a random forest with a drift detector. Drift detection leads to forest pruning to remove tress that have poor performance. Pruned trees are replaced with new weak classifiers [17].

Knowledge Maximized Ensemble (KME) uses a combination of off-the-shelf and created drift detectors to recognize various forms of drift simultaneously. Models are updated when enough training data is collected and removed if they perform poorly [18].

Most methods approach concept drift with an eye towards detection and subsequent normalization. Updating or rebuilding a machine learning model facing drift involves two bottlenecks in the classification pipeline: data labeling and model training; of these, data labeling is the greater challenge due to its oracle requirements. Such wait-and-see models that perform corrections once errors have been detected entail periodic performance degradation before they are corrected with model updates; this may be infeasible in mission-critical applications. Active learning strategies counteract this bottleneck in part; the tradeoff is between highly accurate models and clustered, knowledge-agnostic representations that consider data on distance without subject matter expertise.

3   Data Sources

Our application, LITMUS-adaptive, combines physical sensor and news data, which have high-latency and are scarce, with social sensor data, which have low-latency and are abundant. The social sensor data is used for event detection through machine learning classifiers, while the physical sensors and news data are used to update machine learning classifiers.

3.1   Physical Sensors and News

1. **News**: News articles are downloaded from various online RSS feeds. Each source is described by the article link, the publish date, the article headline, and the publication name. Locations are extracted from the text using Named Entity Recognition. Publisher sources include international feeds from agencies (e.g. BBC, CNN, ABC, Reuters), as well as local news sources (some sample snippets are provided in Fig. 1).
2. **Rainfall Reporting**: We download rainfall data from NOAA and earthquake data from USGS to perform to validate landslide detections.
3. **Landslide Predictions**: The National Oceanic and Atmospheric Administration (NOAA) and USGS provide landslide predictions in select locations where there is enough terrain and rainfall data. LITMUS uses this to perform localized landslide tracking and labeling.

Fig. 1. Snippets of news articles about landslides. Each retrieved article is geolocated using NER to identify locations and indexed spatiotemporally.

3.2   Social Sensors

1. **Twitter**: a keyword streamer is used to download tweets continuously for Twitter. Keywords include the words 'landslide', 'mudslide', and 'rockslide' as well as their lemmas (some examples are provided below).
2. **Facebook**: a general keyword streamer is used to download public Facebook posts. Existing web crawlers are leveraged to improve retrieval efficiency (Fig. 2).

Fig. 2. Sample of raw tweets. The left side are relevant tweets. The right side are irrelevant tweets for landslide detection.

4 Approach

The dataflow for our application is shown in Fig. 3. We perform physical event detection by performing classification on social media data. We use binary classifiers that detect whether a given social sensor post is relevant to a given event or not (e.g. landslides). The latency between a physical event's occurrence and social sensor post about the event is significantly lower than latency with news reports and physical sensors, which often require expert confirmation. In contrast, social sensor data has low latency. However, it lacks the reputability of physical sensors and news reports. We rely on the social sensors for event detection, and physical sensors to continuously tune machine learning models.

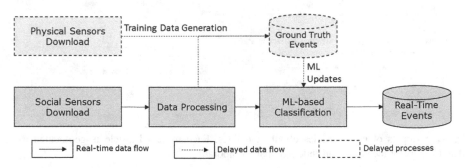

Fig. 3. Our application detects real-time events from social sensors. It remains drift-adaptive by integrating physical sensors with social sensors to continuously update machine learning classifiers.

Traditional approaches perform event detection under a static data assumption. Our contribution is in generating updated machine learning classifiers without any manual intervention so that our application can adapt to concept drift.

4.1   Social Sensor Download

Social sensor download operates in real-time. Our application has streaming endpoints for several short-text social media systems, such as Twitter and Facebook, with an extensible framework for integrating custom streaming sources. Social sensor downloads operate in a high-volume streaming setting. Each downloaded post object contains at least five fields: (i) the post content as a Unicode string, (ii) an array of named locations within the post (this is usually null and is filled during data processing step), (iii) timestamp of post, (iv) array of hyperlink content within the post, and (v) user-id or screen name of human reporter who created the post.

4.2   Physical Sensor and News Download

Physical sensors and news sources are dedicated physical, social, and web sensors providing event information with human annotations. In contrast to social sensors, these are trustworthy sources. We further distinguish physical sensors and news from social sensors: physical sensor and news data is highly structured and contains detailed event information. In our domain (landslides), most physical sensors and news sources provide geographical coordinates and time of landslide disaster. In most mission-critical applications, such physical sensor and news data appears long after an event takes place, once reputable sources have confirmed the event. Additionally, these sensors have lower volume.

For physical sensors, it is trivial to insert the physical event provided by the sensors into our ground truth database by extracting timestamp and location information. News articles provide topic tags that can be mined for an application's event; event reports (e.g. earthquake or large landslide report by USGS) provide detailed information about events, including locations, timestamps, event range, and event impact. We use Named Entity Recognition as well as location tags of news articles to extract location information to identify event location. These events are also stored in the ground truth event database as confirmed landslide events.

4.3   Social Sensor Data Processing

Social sensor data has low-context, which hinders location extraction and classification. Additionally, Named Entity Recognition (NER) often fails on short-text because there are too few words for the NER algorithms. We augment natural language extractors by sharing information between ground truth events and social sensor data processing. We provide an example with location extraction. Since social posts have few words, location extraction is not accurate on the short-text and often misses locations provided in a post's text content. So any location identified by NER is saved in memory for several days as a string. This string is used to augment location extraction with substring match; intuitively, if there is one social post about an event in a location, there could be others. Similarly, any ground-truth event locations are also added to the short-term memory to augment social sensor location extraction.

4.4   Automated Training Data Generation

We integrate social sensor data with physical sensors and news sources to automatically generate labeled data. During live operation, social sensor data is passed to the Machine Learning classifiers for event detection. Concept drift causes performance deterioration in these classifiers. So, our application performs model updates at regularly scheduled intervals. A model update requires labeled data, and a classifier's own labels cannot be used for updating itself. As noted before, it is also impractical to manually label the large volume of social sensor data (on Twitter alone, there are >500M tweets per day).

Our application matches historical social sensor data to ground truth events detected from physical sensors and news reports, which are highly trustworthy. Intuitively, social posts with landslide keywords that have similar space-time coordinates as ground truth events are very likely relevant to a real landslide (as opposed to irrelevant posts such as *election landslides*).

At the end of each data window (one-month windows in our landslide application), ground truth events of the window are stored as cells (coordinates of events are mapped to 2.5-min cell grids on the planet [7]). Social media data from the window is localized by time into 6-day bins. Note that in data processing step, NER augmentation only occurs *forward* in time, i.e. when a new location is added to the memory for substring matching, only subsequent social sensor posts are processed with the new location. So, during the automated training data generation stage, we have access to archived social sensor data.

We take advantage of this by re-processing data from the prior window. Locations extracted from each ground-truth event within ±3 days are used as a substring filter to extract locations from social posts. We then perform automated labeling by matching a social media post's space-time coordinates to true-event location and time. Location matching is achieved with the 2.5-min cell grid superposition using strong supervision. This is in contrast to weak-supervision [21] as our supervisory labeling is domain-specific, instead of domain-agnostic. Table 1 below shows statistics about the generated training data (labeled data) in each window.

Table 1. Automatically labeled data in each data window

Data Window	Data Samples	Labeled
2014-Training Data	26,953	13028
2014-Test Data	6464	3266
July 2018	378	189
August 2018	212	106
September 2018	386	193
October 2018	498	249
November 2018	1770	885
December 2018	446	223

5 Event Detection with Machine Learning Classifiers

Our application uses machine learning classifiers to perform event detection. We employ a variety of statistical and deep learners, including SVMs, Logistic Regression, Decision Trees, and Neural Networks. Our drift adaptation consists of two complementary parts: **Classifier generation** (to create new ML models) and **Classifier updates** (to update older models with new data). We cover them below. We will first describe the classifier generation/update procedure. Then we will cover update schedules which govern when new classifiers are generated or updated.

5.1 Classifier Generation

We define a window as the collection of social sensor data between two updates. At the end of a data window, training data is generated for the previous window using procedures in *Automated Training Data Generation*. The labeled samples are used to train new ML classifiers. Existing classifiers are copied, and the copies are updated with the new data. Both new and updated classifiers are saved to a database using key-value scheme. The classifiers function as values and the training data as the key. Currently, instead of storing the entire training data, we store the training data centroid as the key for a classifier.

5.2 Update Schedule

We support three types of classifier update schedules: *user-specified*, *detector-specified*, and *hybrid*, described below. These schedules allow for continuous classifier generation and updates to combat concept drift.

User-Specified. Users can set up an update schedule (daily, weekly, monthly, etc.). The application tracks the internal time, and when an update is triggered, procedures in *Classifier Generation* are followed to create new classifiers and update existing ones.

Detector-Specified. Some classifiers types provide confidence values with predictions. Neural networks with softmax output layer provide class probabilities. Drift can be detected by tracking frequency of low confidence labels. For linear classifiers (including SVMs), higher density of data points close to the separating hyperplane over time can indicate signal drift. If drift time exceeds a threshold, procedures in *Classifier Generation* are followed to create new classifiers and update existing ones.

5.3 ML-Based Event Detection

Classifiers are retrieved using two approaches: (i) *recency* or (ii) *relevancy*. *Recency* performs lookup on most recently created classifiers. *Relevancy* performs k-NN (k nearest neighbors) search to find training data (using the stored data centroids) that is closest to prediction data; the k-closest data are then used to look up respective classifiers in the classifier database. These classifiers are used as an ensemble).

Our machine learning event detectors use multiple classifiers with votes to perform predictions, since ensembles perform better than lone classifiers. Our ensemble classifier supports several weight assignment options:

Unweighted Average. The class labels predicted by each classifier in the ensemble (*0* for irrelevant and *1* for relevant physical event) are summed and averaged. *Score* ≥ 0.5 indicates majority of classifiers consider the input post as relevant to the physical event.

Weighted Average. Classifiers can be weighted by domain experts based on which algorithm they implement. Weak classifiers (random forests) can be given lower weights than better classifiers (SVMs, neural networks).

Model-Weighted. We can determine classifier weights using their prior performance:

$$w_{M_i^t} = \frac{f_{M_i^{t-1}}}{\sum_a^n f_{M_a^{t_M}}}$$

where M_i is classifier i, $w_{M_i^t}$ is the weight of classifier i in data window t, and $f_{M_i^{t-1}}$ is the validation accuracy of M_i on the testing data in the data window t_M (t_M is the window where the model was last trained).

6   Evaluation of Drift Adaptive Approach

We first demonstrate the need for concept drift adaptive event detection with evidence of concept drift in our data. As shown in Table 1, we have social sensor data from 2014 through 2018. Each labeled post's text is converted to a high-dimensional, numeric representation using word2vec [22]. The post vector is dimensionally reduced with tSNE. The tSNE-based reducer measures pairwise similarities between data points and

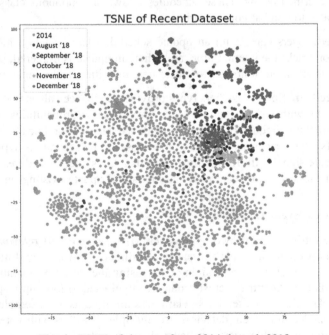

Fig. 4. TSNE of datasets from 2014 through 2018

shows the event characteristics separation between 2014 and 2018 data. Our real-world live data will continue to evolve over time.

Such drift in text data is difficult to predict due to the phenomena of lexical diffusion [23]. Current drift adaptation methods use synthetic data with bounded and predictable drift to evaluate methods; they do not focus on adapting to unbounded drift in live data (Fig. 4).

We evaluate our drift adaptive event detectors with two approaches summarized in Table 2. In each window, the drift adaptive learner is provided with classifiers trained under each approach and tuned with optimal hyperparameters obtained using grid search.

Table 2. Summary of approaches

Approach	Description	Training Data
N_RES (Static)	Non-drift resilient approach with static classifiers	2014 Data
RES (Adaptive)	Drift resilient approach using generated training data for updates	2014 Data – 2018 Data (Separated by windows)

We compare performance of each approach in Table 2 over subsequent windows in our data. As we show in Fig. 5, ensembles with resilience (adaptive approach) outperform non-resilient counterparts (static approach) throughout. RES under both statistical and deep learners maintains high f-score across multiple years. N_RES has significantly higher variance in performance, and is generally poor at adapting to the live data's drift.

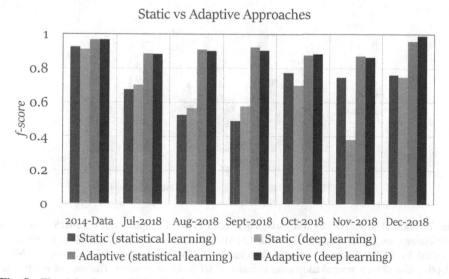

Fig. 5. The Adaptive methods significantly outperform static counterparts in the 2018 data.

N_RES classifiers face deterioration across all metrics without access to generated training data to update their parameters (Fig. 6).

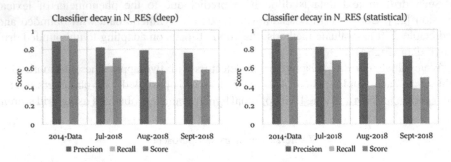

Fig. 6. Decay is apparent in non-resilient classifiers. The f-score is high during the offline window, but degrades without access to generated training data.

7   Landslide Detection: Results

We built our application with physical event detection in mind; evaluation is performed with landslide detection. In the previous section, we validated our drift adaptive approach. Here, we compare our application – LITMUS-adaptive, to LITMUS-static, the traditional approach. Figure 7 shows the raw event comparisons between LITMUS-static and LITMUS-adaptive. LITMUS-adaptive outperforms LITMUS-static, and over time, the share of events detected only by LITMUS-adaptive increases (Fig. 8). We see that by December 2018, LITMUS-adaptive detects 71% of events, compared to LITMUS-static's 28% of events, an increase by 350%. We show global coverage in Fig. 9.

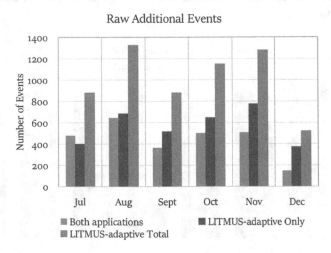

Fig. 7. Raw additional events comparison between LITMUS-adaptive and LITMUS-static. *Both applications* are events detected by both LITMUS-static and LITMUS-adaptive. Every event detected by LITMUS-static was also detected by LITMUS-adaptive. In addition, LITMUS-adaptive also detects several additional events (*LITMUS-adaptive only*). The sum of the two are shown in *LITMUS-adaptive Total*

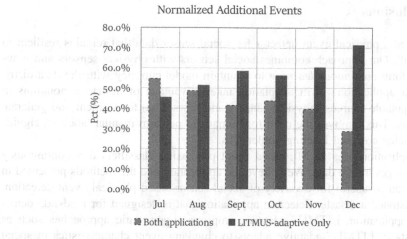

Normalized Additional Events

Fig. 8. Events normalized as fractions of total. Over data windows, fraction of data missed by LITMUS-static increases as its classifiers deteriorate. However, with continuous updates, LITMUS-adaptive can adapt to drift and maintain higher accuracy and thus, better event detection.

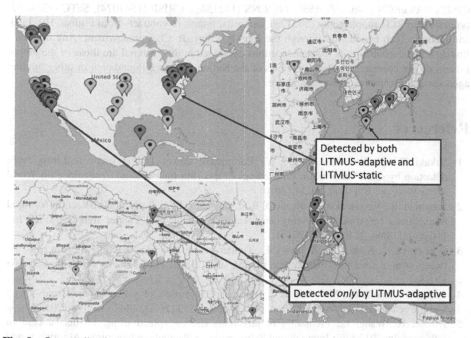

Fig. 9. Our application has global coverage. We are able to detect events across the globe, and as mentioned, every event in LITMUS-static is also detected in LITMUS-adaptive, along with hundreds of additional events LITMUS-static misses (Figs. 7 and 8).

8 Conclusions

We proposed a physical event detector for social sensor data that remains resilient to concept drift. Our approach combines social sensors with physical sensors and news data to perform *continuous learning* to maintain model currency with the data distribution. Our application's drift adaptation takes advantage of human annotations in existing reputable sources (physical sensors and news data) to augment and generate training data. This removes the need for humans to perform manual labeling, significantly reducing cost and labeling bottlenecks.

Our application uses an ML-based event processing classifiers that continuously adapt to changes in live data. We believe the application and the methods presented in this paper can be useful for a variety of social-sensor based physical event detection. We demonstrated a disaster detection application that is designed for landslide detection. Our application, LITMUS-adaptive, improves upon static approaches such as LITMUS-static. LITMUS-adaptive adapts to changing event characteristics in social sources and detects almost 350% more landslide events than LITMUS-static. Moreover, LITMUS-adaptive achieves f-score of 0.988 by December 2018, compared to f-score of 0.762 for static approaches.

Acknowledgement. This research has been partially funded by National Science Foundation by CISE's SAVI/RCN (1402266, 1550379), CNS (1421561), CRISP (1541074), SaTC (1564097) programs, an REU supplement (1545173), and gifts, grants, or contracts from Fujitsu, HP, Intel, and Georgia Tech Foundation through the John P. Imlay, Jr. Chair endowment. Any opinions, findings, and conclusions or recommendations expressed in this material are those of the author (s) and do not necessarily reflect the views of the National Science Foundation or other funding agencies and companies mentioned above.

References

1. Sakaki, T., Okazaki, M., Matsuo, Y.: Earthquake shakes Twitter users: real-time event detection by social sensors. In: Proceedings of the 19th International Conference on World Wide Web, pp. 851–860. ACM (2010)
2. Jongman, B., Wagemaker, J., Romero, B.R., de Perez, E.C.: Early flood detection for rapid humanitarian response: harnessing near real-time satellite and Twitter signals. ISPRS Int. J. Geo-Inf. 4(4), 2246–2266 (2015)
3. Wakamiya, S., Kawai, Y., Aramaki, E.: Twitter-based influenza detection after flu peak via tweets with indirect information: text mining study. JMIR Public Health Surveill. 4(3), e65 (2018)
4. Yang, S., Santillana, M., Kou, S.C.: Accurate estimation of influenza epidemics using Google search data via ARGO. Proc. Natl. Acad. Sci. 112(47), 14473–14478 (2015)
5. Hirose, H., Wang, L.: Prediction of infectious disease spread using Twitter: a case of influenza. In: 2012 Fifth International Symposium on Parallel Architectures, Algorithms and Programming (PAAP), pp. 100–105. IEEE (2012)
6. Musaev, A., Wang, D., Shridhar, S., Pu, C.: Fast text classification using randomized explicit semantic analysis. In: 2015 IEEE International Conference on Information Reuse and Integration (IRI), pp. 364–371. IEEE (2015)

7. Musaev, A., Wang, D., Pu, C.: LITMUS: landslide detection by integrating multiple sources. In: ISCRAM (2014)
8. Thom, D., Bosch, H., Koch, S., Wörner, M., Ertl, T.: Spatiotemporal anomaly detection through visual analysis of geolocated Twitter messages. In: 2012 IEEE Pacific Visualization Symposium (PacificVis), pp. 41–48. IEEE (2012)
9. Lazer, D., Kennedy, R., King, G., Vespignani, A.: The parable of Google Flu: traps in big data analysis. Science **343**(6176), 1203–1205 (2014)
10. Lazer, D., Kennedy, R.: What we can learn from the epic failure of Google Flu Trends. Wired. Conde Nast **10** (2015)
11. Almeida, P.R., Oliveira, L.S., Britto Jr., A.S., Sabourin, R.: Adapting dynamic classifier selection for concept drift. Expert Syst. Appl. **104**, 67–85 (2018)
12. Bach, S.H., Maloof, M.A.: Paired learners for concept drift. In: Eighth IEEE International Conference on 2008 Data Mining, ICDM 2008, pp. 23–32. IEEE (2008)
13. Gama, J., Medas, P., Castillo, G., Rodrigues, P.: Learning with drift detection. In: Bazzan, Ana L.C., Labidi, S. (eds.) SBIA 2004. LNCS (LNAI), vol. 3171, pp. 286–295. Springer, Heidelberg (2004). https://doi.org/10.1007/978-3-540-28645-5_29
14. Göpfert, J.P., Hammer, B., Wersing, H.: Mitigating concept drift via rejection. In: Kůrková, V., Manolopoulos, Y., Hammer, B., Iliadis, L., Maglogiannis, I. (eds.) ICANN 2018. LNCS, vol. 11139, pp. 456–467. Springer, Cham (2018). https://doi.org/10.1007/978-3-030-01418-6_45
15. Markou, M., Singh, S.: Novelty detection: a review—part 1: statistical approaches. Signal Process. **83**(12), 2481–2497 (2003)
16. Lazarescu, M.M., Venkatesh, S., Bui, H.H.: Using multiple windows to track concept drift. Intell. Data Anal. **8**(1), 29–59 (2004)
17. Gomes, H.M., et al.: Adaptive random forests for evolving data stream classification. Mach. Learn. **106**(9–10), 1469–1495 (2017)
18. Ren, S., Liao, B., Zhu, W., Li, K.: Knowledge-maximized ensemble algorithm for different types of concept drift. Inf. Sci. **430**, 261–281 (2018)
19. Shan, J., Zhang, H., Liu, W., Liu, Q.: Online active learning ensemble framework for drifted data streams. IEEE Trans. Neural Netw. Learn. Syst. **30**(2), 486–498 (2018)
20. Žliobaitė, I., Bifet, A., Pfahringer, B., Holmes, G.: Active learning with evolving streaming data. In: Gunopulos, D., Hofmann, T., Malerba, D., Vazirgiannis, M. (eds.) ECML PKDD 2011. LNCS (LNAI), vol. 6913, pp. 597–612. Springer, Heidelberg (2011). https://doi.org/10.1007/978-3-642-23808-6_39
21. Dehghani, M., Zamani, H., Severyn, A., Kamps, J., Croft, W.B.: Neural ranking models with weak supervision. In: Proceedings of the 40th International ACM SIGIR Conference on Research and Development in Information Retrieval 2017, pp. 65–74. ACM (2017)
22. Mikolov, T., Chen, K., Corrado, G., Dean, J., Sutskever, L., Zweig, G.: Word2vec (2013). https://code.google.com/p/word2vec
23. Eisenstein, J., O'Connor, B., Smith, N.A., Xing, E.P.: Diffusion of lexical change in social media. PLoS One **9**(11), e113114 (2014)

ClientNet Cluster an Alternative of Transferring Big Data Files by Use of Mobile Code

Waseem Akhtar Mufti[✉]

Alborg University, Aalborg, Denmark
wmufti@gmail.com

Abstract. Big Data has become a nontrivial problem in the field of business as well as in scientific applications. It becomes more complex with the growth of data and scaling of data entry points. These points refer to the remote and local sources where huge data is generated within tiny slots of time. This may also refer to the end user devices including computers, sensors and wireless gadgets. As far as scientific applications are concerned, for example, Geo Physics applications or real time weather forecast requires heavy data and complex mathematical computations. Such applications generate large chunks of data that needs to transfer it through conventional computer networks. Problem with Big Data applications emerges when heavy amount of data is transferred or downloaded (files or objects) from remote locations. The results drawn in real-time from large data files/sets become obsolete due to the fact data keeps on adding new data into the files and the downloading by remote machines remains slower as compared to file growth. This paper addresses this problem and provides possible solution through ClientNet Cluster of remote computers, Specialized Cluster of Computers, as one of the alternative to deal with real-time data analytics under the hard constraints of network. The idea is moving code, for analytic processing, to the remotely available big size files and returning the results to distributed remote locations. The Big Data file does not need to move around network for uploading or downloading whenever the processing is required from distributed locations.

Keywords: Big data · Mobile code · File transfers · Distributed clients

1 Introduction

1.1 Big Data Sources

Computer is efficient at processing highly complex algorithms. CPU gives a correct and fastest output to the problems if these problems are programmed. The best algorithms available for sorting and searching are still used by many day to day applications involving considerably large amounts of data and higher level of problem complexity. The limitations of computing devices are realized especially when a very large data are processed by a simple program. Problem becomes worse when there are multiple keys given for simultaneous searching of trillions of data items scattered around different

© Springer Nature Switzerland AG 2019
Y. Xia and L.-J. Zhang (Eds.): SERVICES 2019, LNCS 11517, pp. 106–118, 2019.
https://doi.org/10.1007/978-3-030-23381-5_8

data structures or memory locations. Sorting them and then searching infinitely many data items in the scenario when multiple organizations are connected for some common goals of business. The problem of handling Big Data [1–3] is even impossible if the output is required in real-time. The decision making has to be automatic because it is not possible for a human to analyze and assess higher volumes of data in real-time and making business decisions within limited time duration. Geo physical applications involve computations of a number of physical properties of earth layers, wind densities on different altitudes and the computations of tides under water to be the main reason for tsunamis, earth quacks and sea storms.

Initially Big data emerged due to Web logs and machine logs recording for user behavior analysis on internet and offline [8]. The size of web logs evolved with increase in number of computer and internet users, what went to be considered as big volume of data and for the analytics of user behaviors. The volume of big data further exploded as number of internet based applications introduced e.g. social networks, cellular gadgets, E-commerce, cloud computing and all those devices and human sources involved putting inputs to the systems and generate outputs on communication network.

This paper introduces the first version of distributed **ClientNet** cluster developed in Java. It provides the solution as an alternate of transferring very large files to the points of processing. Files are stored on fixed locations whereas processing code, it may be for data analytic purpose, is transferred to the file locations. The code transfer algorithm used in ClientNet is inspired by the visitor design pattern [4] one of the classic software design technique for object oriented software.

1.2 Proposed System

In this paper map computing is implemented in ClientNet cluster system for a given large text files as an example of Big Data. The system consists of 3 clients, 1 coordinator, 1 data server and 1 executor server; all are connected as remote machines. Very large text files to compute the maps are physically stored on the data server. Since the files are large sized therefore do not need to be transported to remote networks for data analytics. Instead, the client codes that can compute the map are transported in parallel to process files on destination node and the results are sent back to the remote clients. This saves network bandwidth, time for real-time data analytics coping with real-time growth of remote data files. The map compute for big files is an example of a scenario to demonstrate mobile code method to deal with Big data. ClientNet distributed cluster is completely developed in Java with built-in RMI and Map compute classes.

The system is in early stage and does not support distributed file system of its own or of any other cluster. It also does not support advanced techniques for in-memory processing of very big data sets which is the focus of my next paper. However, the system provides an efficient coordination among local hosts simulating real scenario of remote map compute job. The system is flexible enough that clients can add as many jobs as they want by adding Java classes for each job. All client jobs execute in parallel using multithreaded Java model of concurrency. It has used the power of Java classes and objects that provides an early version for in-memory processing avoiding the frequent read/write accesses on physical storage. The system is less complex and is tailored directly to serve the processing of remote jobs as compared to Apache Hadoop [5] and Apache Spark [6]. ClientNet is platform independent and fully demonstrates independent functioning of all components of clustered computers distributed over remote locations. This paper is composed of Big Data concepts, its architectures and challenges; introduces the first version of ClientNet cluster system as programming solution for Big Data transfers and data analytics by map computing as an example, its design model and concurrency model.

2 Big Data Concepts and Architectures

2.1 Foundations

According to O'Relly Media [7], who first coined the word Big Data in 2005, they define it as *"Big data is data that exceeds the processing capacity of conventional database systems. The data is two big, moves too fast, or doesn't fit the structures of your database architectures. To gain value from this data you must chose an alternative way to process it"*. Almost everywhere in the literature the size of big data is not termed as fixed to designate it to be Big Data. I would consider a big data when it is not possible to handle it with traditional relational database tools and techniques. This includes the size of data must be large enough that cannot be accommodated in database tables, or unstructured enough to extract its meaning, or a continuously growing data that is not possible to be placed in a database container of fixed size. Big data can be considered if it is large enough that available searching and sorting algorithms cannot be applied as they are used for conventional systems. This intrinsically poses the possibility of creation of new techniques and algorithms to target for the typical nature of data. For example, if data is large enough that cannot be passed to remote computers (one of the main focus of this paper) or the processing is possible only through data mining techniques that lead to data clusters and mapping techniques.

More specifically Big data is characterized based on its specific properties known as V's of Big data [1, 2]. Since this paper is not dedicated to the survey and detailed definitions of big data and its available clusters, therefore I have limited its text to focus on the contribution of paper in addition to the brief descriptions.

Volume: It refers to the size that would be equal or beyond the maximum of its size present at the time of writing this article. The volume may possibly go beyond of multiples of petabytes. While considering the distributed big data scenario then it is possible the volume would cross thousands of petabytes. This would lead it to infinitely big enough to measure the size of continuously growing files and the only way to consider it would be by using partitioning algorithms. One can imagine the difficulties of searching and sorting that would need specialized and context bound techniques to achieve goals for every special scenario.

Variety: As given above one of the distinct factors of big data is the variety of data due to which it requires advanced techniques for processing. Types of data may be unstructured data generated through social media in form of random tweets and file attachments produced in multiple contexts of conversation. Unstructured or unformatted data normally cannot be used if data filtering is not applied. However the other formats of data e.g. media files: videos, images, different texts: doc, xml, pdf, txt, etc. are not difficult to maintain in available databases. In this case if the data is continuously being generated to build enormously big size then new techniques would require for real-time data analytics.

Velocity: Rate of growth of data is velocity which is the most crucial factor to deal. This is the biggest challenge that has pushed computing professionals to device new algorithms and high capacity storage devices and high speed computer networks.

Veracity: It is the incomplete or noisy data that makes analytics more difficult. Data is periodically monitored preventing it garbage data. For this purpose data filtering is applied or manually the developers filter it before analytic process begins.

Validity: This refers data must be valid and consistently available in a distributed system of computers. The replicated data must be taken care of its validity before extracting its meaning at different locations in real-time. If data is not valid at all points of processing then the extracted meaning would not be valid as well and results would be inconsistent. This is also the basic property of conventional databases.

Volatility: This is one of the difficult tasks in big data that continuously keeps evolving into different volumes and variety. It refers to the data that is no longer relevant must be discarded or not to store it in valid data containers. To save the space from unnecessary data the volatility processes must be monitored continuously as garbage collection is performed in different systems and languages.

2.2   Big Data Architectures and Technologies

After that the paper moves on the actual goal defined in the title. Big data architecture framework [9] defines several of its models, formats, management methods, analytics methods, infrastructure (storage, methods of accessing, processing and routing of data) and security. Big data technology includes programming tools that provide the solution to the big data architecture framework. This involves technological framework e.g. Hadoop and Spark clusters which are also called big data architectures. These tools are the collection of several components that provide data processing, analytics, storage and distribution solution along with powerful compute engine. The overall data life-cycle and the architecture is called big data ecosystem [9–11]. These large scale clusters can process big data in parallel through unified framework of components for each service. For example, Hadoop contains Hadoop Distributed File System (HDFS) for managing very large data files into partitions spanned over thousands of remote nodes, MapReduce, Mahout, HBase, OoZie, Pig, Flume, Zookeeper, Hive, Sqoop, Whirr, etc. As shown into the well known diagram these tools are integrated in Hadoop tool framework (Fig. 1).

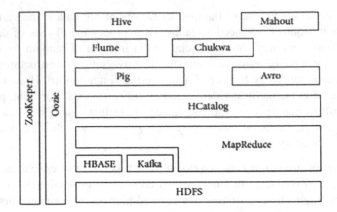

Fig. 1. Hadoop tool suite [12]

One of the famously known tools is MapReduce which is collection of programs written in Java and Scala. This system is used to transform raw text data into counted words from collection of text files so that the output can be queried or analyzed automatically; reduces the text into words and their occurrence as given in the following diagram. Originally MapReduce was developed by Google since then it has been used by others as well. This is one of the big data analytic tools provided by collection of Java classes used in this paper as an example to demonstrate ClientNet cluster. So far it is the solution used only for text processing on large scale distributed systems. For instance it counts words or users and their behavior on Web. Further details are given after the section of Spark.

Mahout is developed by Apache used for distributed collaborative filtering, clustering and classification of data. It is written in Java and Scala therefore supports both languages. The latest version of Mahout is a distributed linear algebra framework designed to let mathematicians, statisticians and data scientists quickly implement their own algorithms (Fig. 2).

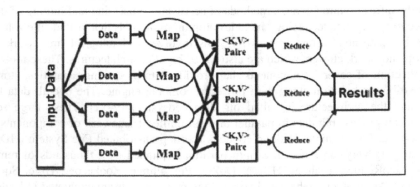

Fig. 2. MapReduce [13]

Microsoft's Azur HDInsight [14] was initially known as Windows Azure uses popular open source frameworks including Apache Hadoop, Spark and Kafka. Azur is cost-effective and provides enterprise-grade service for open analytics. It integrates seamlessly with components of open source echo system of above mentioned clusters with global scale. *"It is a cloud computing platform, designed by Microsoft to successfully build, deploy and manage applications and services through a global network of datacenters"* [15].

For the purpose of simplicity and focus I have deferred more details on other clusters such as Spark, one of highly significant big data clusters. It is more famously known for its in-memory computing and scheduling of availability of large memory objects. The most significant work of Spark is it supports both object oriented programming and functional.

3 ClientNet Cluster

3.1 Introduction

ClientNet is set of Java programs running on a cluster of computers connected by communication network. The first version of it contains 3 clients, 1 coordinator, 1 data server and 1 executor engine. It is scalable to a number of similar clusters providing solution to remote clients by cascading the design to multiple physical locations. Since it is non-commercial and in the early stage of its development therefore currently the ClientNet is simple and limited to minimum number of computers communicating via passing messages through Java Remote Method Invocation protocol as shown in the following diagram. The messages are passed by invoking remote functions and passing objects as parameters.

To keep it simple and working I have implemented the map compute using built-in Java **HashMap** [16] for a large size text files that takes long time if transmitted to remote computers. Taking the advantage of message passing, the computing code that is responsible of map analytics is passed from parallel clients to the remote data server where big size text file is physically available. Since the map compute code is light weight therefore it is easier to pass around remote network and execute at the remote server side. After completion, the analysis results (map entries) are sent back to the remote clients as Java objects which are lighter enough for smooth data communication. The process involves coordination of all computers running Java clients and server programs. Since the Java objects are memory residents therefore map compute does not executes frequent reads and writes on hard drive.

3.2 ClientNet Architecture and Programming Model

The cluster system consists of a group of computers coordinating via programming model based on Java RMI [17] as given in the following (Fig. 3).

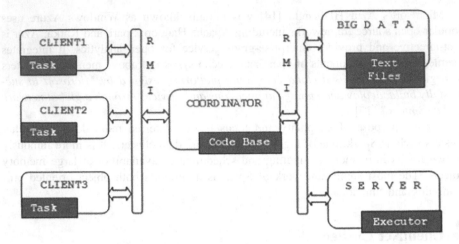

Fig. 3. ClientNet cluster connections diagram.

3.2.1 Connectivity and Message Passing

All machines are running Microsoft Windows operating system. The next release of ClientNet would adopt heterogynous platform. Each computer participating in cluster needs to get connected on communication network before it runs any program. For this purpose each computer registers to RMI registry with computer name recognizable on the network and its port number where it receives and sends messages to other computers as given in the following Java code:

```
Registry DSregistry = LocateRegistry.createRegistry(1099);

DSregistry.rebind("Data", new Data());
```

The RMI registry comes with Java distribution and it must execute on each machine with JVM before the start of any Java Client-Server program. The computer name, e.g. "**Data**", is name string that would be globally known to all of computers in the cluster. Since the big text file does not transfers to other computers therefore Data computer does not send messages to other computers rather it only receives requests from Coordinator for file read access. Once the machine is available on the network it publishes its I.P address, port number and name string to show its global availability. When the Client1 machine, for instance, sends messages to other computer in the cluster group it executes the following lines of Java code before actual message passing.

```
DSregistry = LocateRegistry.getRegistry(6000);

coordinator = (CoordinatorInteface) DSregistry.lookup("127.0.0.1");
```

This means **Client1** looks for other computer's I.P address and its port number by accessing RMI registry function `lookup()`; Here it finds **Coordinator** machine of port number 6000 and localhost identified by I.P: 127.0.0.1. After creating link the

Client1 creates **reference** named `coordinator` to the Coordinator's remote object on Client1 local disk. Since it has Coordinator's reference therefore it can send and receive messages to remotely available Coordinator by invoking interface functions on `coordinator` as given below:

```
plainTextFile = coordinator.getTextFile();
```

The `getTextFile()` is remote interface function of remote Coordinator. The class diagram of ClientNet cluster is given in the following Fig. 6. This function returns the remote reference, to the Coordinator, of type `File` of the big data text file that resides on remote Data server. For security purpose the Coordinator represents Data server for all clients. Everywhere in the cluster the communication takes place among all computers as discussed above example. Remote references are first obtained after creating RMI connections.

3.2.2   Data Transmission and Filtering

Each client is a Java class contains inner class **WordCount** which contains a thread that runs anywhere where it is invoked at destination. The inner class instantiates map object of type `HashMap`. It is built-in Java utility for map computing jobs, records a count for each word entry e.g. `map.put(word, 1)`. Here the `put()` function invokes on **map** object and receives two parameters i.e. **word** text token and its occurrence **1**. The `HashMap` class does not provide filtering functionality by default therefore I have added extra function `applyfiterWord()` that filters out each scanned word from 5 MB text file and truncates the special characters attached to the word. For instance if the given text is:

"**Among the highest living standard cities of the world are Switzerland, Oslo and Copenhagen etc. Switzerland is the most beautiful as compared to Oslo or Copenhagen.**"

The returned map of **ClientNet** program after applying filter function is: {Switzerland = 2, Copenhagen = 2, Oslo = 2, ...}, where as the original Java map would return it like this: {Switzerland = 1, Copenhagen = 1, Switzerland, = 1, Oslo = 2, Copenhagen. = 1, ...}. The original Java map computes it as: "**Switzerland,**" and "**Switzerland**" two different words and "**Copenhagen**" and "**Copenhagen.**" as two different words. It is because Java map includes last punctuation marks (, and .) attached to the word as the part of that word. The function `applyfilterWord()` truncates the punctuations marks attached and considers it the same word if the word has appeared before.

Therefore filter function prevents and it does not count the new word occurrence. Each time a word is added after filtering unless the complete text is processed. The text filtering function is lengthy enough to present in this paper therefore for clarity purpose one of its checking conditions is provided in the following diagram. This condition filters for surrounded words, e.g.: "`(Apple)`..." Or "`[Apple]`..." Or

"(Apple, Orange)..." and generates only two tokens "**Apple**" and "**Orange**" as two distinct clean words.

The following peace of code given in Fig. 4 is one of the several **if** conditions that filters out some of the special characters and returns a clean word that becomes part of the map entries. The condition checks if the special characters occur at first and last position of a word, that kind of word is a surrounded word e.g. "(Apple)" that is surrounded by two round braces. This condition may further be extended by adding more checks of all special characters. After the condition is evaluated, program copies characters other than braces into another array **surroundedTail** and finally converts into String object **cleanWord**. The program segment is part of a loop which continues for all scanned words.

```
  if(chars.length > 1){
      if((chars[0]     ==     '*'    &&     chars[chars.length-1]    ==
'*')||(chars[0]     ==     '('    &&     chars[chars.length-1]    ==
')')||(chars[0]     ==     '['    &&     chars[chars.length-1]    ==
']')||(chars[0]     ==     '{'    &&     chars[chars.length-1]    ==
'}')||(chars[0]     ==     '\''   &&     chars[chars.length-1]    ==
'\'')||(chars[0]    ==     '-'    &&    chars[chars.length-1]    ==    '-
')||(chars[0] == '<' && chars[chars.length-1] == '>')||(chars[0]
== '(' && chars[chars.length-1] == ',')||(chars[0]    ==    '[' &&
chars[chars.length-1]    ==     ',')||(chars[0]     ==     '{'    &&
chars[chars.length-1] == ',')){
char[] surrounded = new char[chars.length-2];
for(int x=1; x < chars.length-1; x++){
     surrounded[x-1] = chars[x];
     cleanWord = new String(surrounded);
  }
  return cleanWord;
  }
}
```

Fig. 4. One of the filtering condition taken from the function `applyfilterWord()`.

The ClientNet does not support shuffling of intermediate map results because at this stage it does not support distributed file system which has been left for next version. The system assumes final piece of file and begins map compute when it receives the file. All three clients contain the similar functionality which can be changed at anytime with new data analytics functions. It should be noted that data analytics is not the main focus of this paper.

After that the Coordinator packs all the clients' source codes as binary code in form of objects into the Java array list data structure which is called **Code Base**. The code base contains all threads ready to execute wherever the array list is to be accessed. The code base (array lists) is then transferred from Coordinator to Executor server through

RMI protocol. As the number of client requests will increase the size of code base will also increase. The following peace of code as given in the Fig. 5 shows how the Executor runs each client and finally traverses the list and executes each client thread by the function **startClientThread()**.

The Executor program segment traverses the list, accesses each of its remote objects and type casts each object to **ClientCommonInterface** by creating local reference **client** to the related remote Client object. Once the object reference is created it is then used to invoke remote function on it. The invoked function executes the embedded thread of inner class object of the remote client; this is just to remind that data analytic jobs are embedded into the inner class of each Client.

```
    private void runClients() throws InterruptedException, Re-
moteException{

    int x = 0;
    // process remote clients
    while(x < coordinator.getCollection().size()){

        client     =     (ClientCommonInterface)     coordina-
tor.getCollection().get(x++);

    client.startClientThread();
    gc();
        }
    }
```

Fig. 5. Processing of clients threads at executor server

It is assumed that Coordinator and Executor Server are physically near the location of Data server. Therefore for clients the data access becomes easier as compared if the big files would have been accessed by remote clients independently. The final **computed map** is small size text file that contains map entries. This file can easily be sent to anywhere for knowledge extraction or can be transformed into transactional data by inserting into conventional database tables for query processing.

3.2.3 ClientNet Java Classes

The **class diagram** of full system is given in the following Fig. 6. All clients have similar connections as shown in the diagram which shows only one client. All classes are connected through their related RMI interfaces which are shown on top of the class diagram. Since the text file is accessed only for reading purposes therefore there is no concurrency issues are raised. However all clients are parallel and disjoint because of each one is remotely available and physically independent. The **plus** connector of Client1 shows its relation with its inner class WordCount. The Client1 class is shown empty because analytic jobs are nothing but the inner classes of their related client classes. In this paper only one job is given which is embedded into single inner class. All classes and interfaces boxes contain functions written in their body. The **diamond** connector means the reference to remote object and the normal arrow means inheritance among classes or implementation of interfaces. Class diagram may become more complex if more jobs are added. The sequence in which the system must execute is as follows: First of all run the **start** (connectivity) programs of the Data server, then all Clients, then the Coordinator. Finally Executor server runs its application client that executes its processing engine.

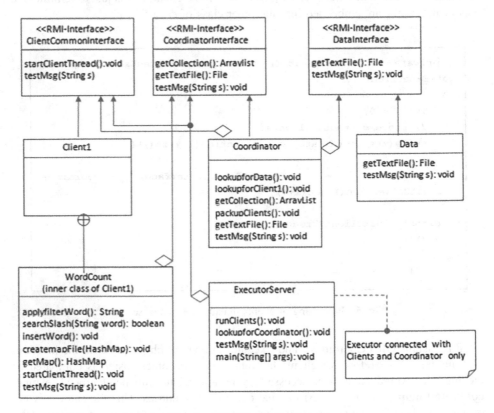

Fig. 6. ClientNet UML classes

At this stage ClientNet is light weight and allows limited functionality of map compute. The uniqueness of this cluster is that the data analytic code, for instance map compute, does not need to be available on the node; rather the node would be passed the source code when analytics is required. This allows light weight and less complex nodes. Therefore every time the data analytics would be performed as a client request. This will allow parallel processing of analytics on loosely coupled resources. The mobile code would enhance mobility of soft resources within the cluster as it is naturally provided by the cloud architecture. It does not support or use any available distributed file system and also does not scales over multiple servers to provided availability of remote resources. I believe if the system can successfully show the basic functioning it can easily be scaled on a number of machines with advanced analytics functionality. The most difficult tasks that would arise are possibly memory management of large objects, scheduling of remotely available storage and processing resources.

4 Conclusion

The cluster successfully implements mobile code technique in Java. ClientNet cluster is scalable to number of clients distributed across remote locations. More work is required to add data analytics of different domains, memory management for large number of objects of big size. Distributed file system and the availability of remote processing nodes is yet to be developed in its next versions. Since the data analytics is not covered. ClientNet can also successfully perform concurrent business transactions with accurate number computation from remote clients in an online scenario. This functioning of the cluster is not present in this paper because banking data transactions is not the subject of this paper. The transactions component of ClientNet cluster implements mutual exclusion to obtain correct calculations by using many of built-in concurrent data structures and functions of Java.

References

1. Watson, H.J.: Tutorial: big data analytics: concepts, technologies and applications. Commun. Assoc. Inf. Syst. **34**, Article no. 65 (2014)
2. Mary, A.J., Arockiam, L.: A study on basic concepts of big data. Int. J. Emerg. Trends Comput. Commun. Technol. **1**, Article no. 3 (2015)
3. Wang, Y., Kung, L.A., Byrd, T.A.: Technol. Forecast. Soc. Change **126**, 3–13 (2018)
4. Gamma, E., Helm, R., Johnson, R., Vlissides, J.: Design Patterns: Elements of Reusable Object-Oriented Software, 1st edn. Addison-Wesley Professional, Boston (1995)
5. Polato, I., Goldman, A., Re, R., Kon, F.: A comprehensive view of Hadoop research – a systematic literature review. J. Netw. Comput. Appl. **46**, 1–25 (2014)
6. Apache Software Foundation. Apache Spark Survey 2016 Report, DATABRICKS (2016)
7. Dontha, R.: Big Data. www.digitltransformationpro.com
8. Mohanty, H.: Big data: an introduction. In: Mohanty, H., Bhuyan, P., Chenthati, D. (eds.) Big Data. SBD, vol. 11, pp. 1–28. Springer, New Delhi (2015). https://doi.org/10.1007/978-81-322-2494-5_1
9. Demchenko, Y., Membrey, P.: Defining architecture components of the Big Data Ecosystem. In: International Conference on Collaboration Technologies and Systems (CTS) 2014. IEEE, Minneapolis, MN, USA (2014)
10. Oussous, A., Benjelloun, F.-Z., Lahcen, A.A., Belfkih, S.: Big data technologies: a survey. J. King Saud Univ. Comput. Inf. Sci. **30**, 431–448 (2018)
11. Joseph, C.P., Thulasi, B.S., Susmitha, V.: Big data – concepts, analytics, architectures – overview. Int. J. Eng. Technol. (IRJET) **5**(2), 125–129 (2018)
12. Khan, N., et al.: Big data: surveys, technologies, opportunities, and challenges. Sci. World J. **2014**, 18 (2014)

13. Zerhari, B., Mouline, S., Lahcen, AA.: Big data clustering: algorithms and challenges. In: International Conference on Big Data, Cloud and Applications BDCA 2015, Morocco (2015)
14. https://azure.microsoft.com/en-us/services/hdinsight/
15. Microsoft Azure Tutorial. www.tutorialspoint.com
16. https://www.javatpoint.com/java-hashmap
17. https://www.javatpoint.com/RMI

Utilizing Blockchain Technology to Enhance Halal Integrity: The Perspectives of Halal Certification Bodies

Marco Tieman, Mohd Ridzuan Darun[✉], Yudi Fernando, and Abu Bakar Ngah

Faculty of Industrial Management, Universiti Malaysia Pahang,
Lebuhraya Tun Razak, 26300 Gambang, Kuantan, Pahang, Malaysia
mridzuand@ump.edu.my

Abstract. Brand owners face multiple challenges in establishing end-to-end halal supply chains and in halal issue management. To this end, leading Halal Certification Bodies (HCBs) set up a discussion group to achieve consensus on a way forward and to examine the potential role of the halal blockchain in resolving these issues, as well as the key parameters, and segregation and communication requirements of the blockchain. Halal issues can be divided broadly into three areas: contamination, non-compliance and perception. Only cases involving contamination and non-compliance need to be reported to the HCB. Consensus has been achieved in the segregation of halal supply chains in terms of designated halal transport, storage and halal compliant terminals, for Muslim (majority) countries, whereas in non-Muslim (majority) countries greater leniency is possible. Effective segregation is only possible with effective communication, whereby the term 'halal supply chain' is encoded in freight documents, on freight labels and within the ICT system.

Keywords: Halal supply chain · Blockchain technology · Halal Certification Bodies (HCBs)

1 Introduction

Halal supply chains have inherent problems or flaws, namely in traceability (the ability to examine and reproduce the exact path of a product in the supply chain) and in organising product recalls; transportation and warehousing (storage) downstream compliance with halal requirements; end-to-end chain integrity (unbroken chain) from source to point of consumer purchase; different halal systems and interpretations of different markets; and lack of IT integration with supply chain partners and halal certification bodies (HCB). A series of high profile halal crises in recent years involving well-known brands have shaken public confidence in the ability of brand owners to assure the authenticity of halal certified products.

Halal certification is dependant upon the compliance of the production process and ingredients used with Shariah requirements. As a result, current halal standards provide product certification, but not supply chain certification [1], and HCBs have limited insight into the level of conformance of the end-to-end halal supply chain (network). Whenever a

Y. Xia and L.-J. Zhang (Eds.): SERVICES 2019, LNCS 11517, pp. 119–128, 2019.
https://doi.org/10.1007/978-3-030-23381-5_9

halal issue arises, a rigorous investigation of the supply chain is needed, not only by the brand owner itself but also by the HCB that certified the product. These investigations may be lengthy, and create uncertainty among Muslim consumers resulting in depleted sales and damage to the reputation of the brand owner concerned. Hence, a systematic database is required to assist respective parties to undertake the investigation process.

Utilising blockchain technology ensures the notification of all completed halal supply chain transactions to all parties within the chain, including the end consumer. A halal blockchain is envisioned to be a private blockchain (different from a public version), whereby the administrator (the brand owner) can determine the rights of each category of supply chain participant, including what is visible and what information can be added to the blockchain. These supply chain transactions are constantly growing as 'completed' blocks are added with each new set of transactions. The blocks are added to the blockchain in a linear, chronological order with a timestamp. Each node in the halal supply chain network gets a copy of the blockchain, which gets downloaded automatically upon incorporation into the network. The halal blockchain has complete information about the participants' addresses and their supply chain path right from source to the point of consumer purchase. As the halal blockchain is shared by all nodes participating in a halal supply chain network, information is easily verified by scanning the QR-code (a two-dimensional barcode) on a product.

Halal blockchains pre-program the requirements of the destination market, such as halal production certificate requirements, halal storage-transportation-terminal handling terms, the encoding of 'halal' on freight documents. Furthermore, these blockchains facilitate in managing halal issues and crises. They also facilitate in organising product recalls, for both trades (so-called silent recalls) and the public (broadcast to the consumers). In a blockchain you can easily identify the parties that have committed fraud, as this information remains visible, thus discouraging companies in a halal supply chain from fraudulent action. At the same time it enables the rating of halal logistics service providers, distributors, and other supply chain participants based on the performance of their services [2].

Increasing the transparency of halal supply chain activities would promote greater consumer trust in the authenticity of a halal brand. Hence, blockchain technology could be the means to enhance this transparency since it offers distributed ledgers and smart contracts. This paper discusses the important findings of a large discussion group (involving leading HCBs) on key issues in global halal supply chains leading to the proposal of the blockchain system as a solution. The rest of the paper is divided into four sections covering the context of using blockchain technology in halal supply chains, the appropriate methodology for a large discussion group, the results of the discussions, and the conclusion.

2 The Context of Blockchain Technology in Halal Supply Chains

The integrity of halal supply chains is not only of great concern to the Muslim consumer, but also to brand owners in protecting their reputation as providers of halal products [3]. Currently, there are insufficient technologies to address these halal supply chain risks. Mohamed et al. [4] suggest that the integrity of halal food must be monitored so that consumers are satisfied with the authenticity of halal products. Consumers can be expected to ask not only for halal products, but also for guarantees that all the activities involved in the supply chain process are conducted in accordance with halal requirements [5, 6].

To remain competitive in the halal industry, Mohamed et al. [4] propose an effective traceability system to allow the stakeholders to detect and track vital information at each stage of the supply chain. Such a system would ensure a reduction in the number of product recalls and non-compliance cases and enhance the halal status of the product. This view is supported by Zailani et al. [7] who believe that a good traceability system would mitigate the risks associated with halal products, as well as functions as a tool for communication, making information available along the supply chain. A transparent and credible halal supply chain depends on a reliable traceability system and cooperation from all partners.

According to Zailani et al. [7], a halal traceability system acts as a tool for communication and makes information available across the supply chain network. Therefore, a variety of IT tools need to be in place to make the system more effective. With the proper technologies, relevant information can be traced back to its source leading to the identification of the point where the halal violation occurred. A halal traceability system requires a proper IT system with appropriate hardware, software and network equipment. Current technologies use a combination of RFID and barcode or QR code [7, 8]. However, there are currently no real time halal tracking systems on the market. Even though tools are available, many studies agree that existing systems are not fully automated, susceptible to fraud, not fully secured, not supported by real time tracking, slow to function, and not available to all consumers [7, 9].

Even though existing technology is able to track and trace goods from suppliers to retailers, there are still many problem cases every year, and these have a major impact on the sales and reputation of businesses. Such problems are inevitably caused by the difficulties in running complex halal supply chains. One of the potential solutions is to use technology to enhance the transparency of the whole supply chain. One emerging technology frequently mentioned nowadays is the Blockchain [10, 11]: notably, it has the potential to solve some of the integrity issues in supply chains [2]. For example, IBM and Walmart [12, 13], conducted a pilot study using blockchain technology to track food products at every stage from the farm all the way to the store in order to improve the confidence of Chinese consumers in food quality. The results show that the blockchain system has the ability to improve food traceability as well as the transparency of the supply chain operations. Halal blockchains have therefore the potential to simplify the management of halal supply chains, not only for the brand owner but also for the HCBs.

3   Methodology

Universiti Malaysia Pahang (UMP) is currently undertaking a research project on halal blockchains to develop a prototype and test the application of blockchain technology for halal food, cosmetics, and pharmaceutical supply chains. The objective of the large discussion group is to determine the parameters of the halal blockchain that are determined by the HCBs, contributing to effective halal issue and crisis management as well as an effective alignment of the halal supply chain with the requirements of the destination market. In particular, this research intends to determine the halal issue scenarios and role of the halal blockchain in these scenarios; the halal supply chain key parameters for the blockchain; and the halal supply chain segregation and communication requirements.

3.1   Organisation of the Sessions

The objective of this large discussion group was to obtain consensus [14] on the: (1) halal issue scenarios in halal supply chains; (2) the role of the halal blockchain in each halal issue scenarios; (3) halal supply chain parameters for the halal blockchain; and (4) halal supply chain segregation and communication requirements for both Muslim and non-Muslim countries.

3.2   Recruiting the Participants

For the discussion group, UMP invited foreign halal certification bodies & authorities recognised by the Malaysian halal authority Jakim to participate [15] as well as additional bodies approved by the Indonesian halal authority MUI [16].

3.3   Conducting the Discussion Sessions

The discussion group session was held in Kuala Lumpur (Malaysia) on 14 December 2017. It was attended by 12 participants and was conducted during one afternoon, allowing HCBs from Asia, Europe and the Middle East to participate via conference call. The discussion comprised presentations on the current issues with halal supply chains, blockchain technology, the halal innovation project, and the functionalities of the halal blockchain based on the results of earlier focus group sessions within the industry.

3.4   Analysing and Reporting

The group discussion was recorded and then transcribed [17, 18]. In line with Ruyter [19], Walden [20] and Chambers and Munoz [21], proposals were classified into categories in order to establish reoccurring themes and patterns [18]. The validity, correctness or credibility of a large discussion group session can be identified through a variety of strategies [22]. Prince and Davies [23] have identified moderator bias as a serious concern in conducting focus groups that can involve the content, the process or participation and the interpretation of the research results. According to Grudens-Schuck et al. [18], questions should be arranged from general to specific to invite openness and avoid bias. In addition, as argued by Prince and Davies [23], the moderator (the researcher of this study) should be well versed in the topic of halal supply chain management, which has been the case through the moderator 12 years of experience in conducting academic research on halal supply chain management and consultancy projects on this same topic with the halal industry. Wall [24] argues that the representativeness of the participants is an issue in large discussion groups. This issue was anticipated by the researchers who requested UMP to base the invitations for the discussion group on the HCB lists from Jakim and MUI.

4 Results of Large Discussion Group

As stated by Heugens et al. [25], a halal issue can be defined as a gap between the stake-holder's expected and perceived halal practices of a brand owner. The participants of the discussion group uniformly agreed that halal issues could be categorised into three different problematic areas as shown in Table 1. The three areas are contamination, non-compliance, and perception. A halal issue could first of all be related to contamination. A Muslim consumer may be uncertain whether he or she has been 'poisoned' with haram (forbidden) ingredients in the food, cosmetics, pharmaceuticals or other halal product purchased. Can he or she still trust the purity of the product and brand? Secondly, there could be an issue of non-compliance through an increased risk of contamination of the product with non-halal elements due to breakages in the halal supply chain as well as issues with the halal certificate or logo. This scenario would make the halal status of the product doubtful, and according to their religion Muslims should avoid a product when in doubt [26, 27]. Thirdly, a halal issue could also be categorized as perception, where there is a possible mismatch between the perceived brand image and the Islamic way of life [28]. For example, the brand owner may use a dubious advertising campaign that could potentially lead to a negative perception of the product. The group agreed that, there should be no requirement for the blockchain system to provide information to the HCB relating to perception issues.

Table 1. Halal issues.

1	Contamination	Example
a	Conterfeit product with halal logo	Conterfeit milk power with halal logo
b	Haram ingredient or contamination with haram	Claim of contamination of halal certified chocolate with alcohol or pork
c	Poisonous ingredient in halal certified product	Mercury in skin whitening cream which is halal certified
2	Non-compliance	Example
a	Expired halal certificate	Expired halal certificate of product, ingredient, or restaurant outlet
b	HCB of ingredient is not recognised anymore	On the recently updated list of recognised HCBs of Malaysia's halal authority JAKIM, a formerly recognised German HCBs is removed from this list
c	New fatwa making halal certified product haram	New religious ruling on banning certain slaughtering methods

(continued)

Table 1. (*continued*)

2	Non-compliance	Example
d	Non-compliance in logistics	Mixing of halal certified product with beer/pork products in logistics or curtain sider with beer/pork/sexy lady
e	Fake or wrong halal logo on product or outlet	In Europe fake halal logos are used on products and restaurants as there is no law against it
f	Fake halal certificate	The manufacturer or restaurant carries a fake halal certificate
3	Perception	Example
a	Wrong commercial/advertising	Sexy dressed lady in commercial
b	Wrong promotion/co-packing	Saving pig or product containing alcohol co-packed with halal certified product
c	Brand image of company is not positive	Company is boycotted because it is an American company
d	Halal authenticity is questioned of company	Company is involved in child labour, pollution, etc.
e	Halal issue that has been resolved already by HCB	A halal issue goes viral (again) which has already been resolved by the HCB

The discussion group uniformly agreed on what information is required by the HCB for an investigation of halal issues (see Table 2). When the brand owner is faced with a halal issue, the blockchain could assist the brand owner by immediately sharing the necessary blockchain information to the HCB so that the investigation time needed can be reduced to a minimum.

Regarding the key halal supply chain parameters, there are three of utmost importance, namely the demographics of the destination country (Muslim majority country or non-Muslim majority country), the product characteristics and the product environment. The demographic destination country is of particular importance as local HCBs base their halal requirements on their Islamic ideology, religious rulings (fatwas) and local customs. The products produced for and exported/distributed to these markets should comply with the local requirements set by the local HCB of the halal supply chain destination market.

In line with Tieman et al. [29] and IHI Alliance [30], the HCBs recognized the importance of differentiating between Muslim (majority) countries, where halal and non-halal products can be more easily distinguished, and non-Muslim countries, where this is often not the case. It has been argued by HCBs that halal supply chains for non-Muslim countries should be less stringent and more practical given the complexity of non-Muslim countries, and that they should not create unnecessary hardship for both Industry and Muslim consumers ensuring the availability of halal products in non-Muslim countries as the most important goal while still achieving an acceptable level of segregation.

Table 2. Information requirements by HCBs with a halal issue.

	Information	Contamination scenario	Non-Compliance scenario	Perception scenario
1	Halal issue scenario	Yes	Yes	No
2	Actions undertaken by company	Yes	Yes	No
3	Supply chain structure	Yes	Yes	No
4	Chain of custody	Yes	Yes	No
5	Product information:	Yes	Yes	No
a	Raw materials	Product name & batch number, HCB, expiry date, lab record	Product name & batch number, HCB, expiry date, lab record	
b	Product	Product name & batch number, production record, production recipe, HCB, expiry date, lab record	Product name & batch number, HCB, expiry date, lab record	
c	Restaurant (in case relevant)	Outlet name, production record, product recipe, HCB, expiry date, lab record	Outlet name, HCB, expiry date	
d	Other supply chain party (in case relevant)	None	Scope, HCB, expiry date	

Furthermore, as also stated by Tieman et al. [29] and IHI Alliance [30] there is consensus among HCBs regarding the sensitivity of bulk (not packed) versus unitized (packed) products as well as the sensitivity of cool chains (closed circulation of air and moisture development) versus ambient environments (dry). Based on the above parameters, full consensus was achieved on the proposal to segregate halal supply chains for transportation, storage and sea/air terminals as shown in Fig. 1.

The discussion group also agreed to standardize communication in relation to freight documents, freight labels and ICT systems in order to ensure the effective identification of halal cargo in international trade. In line with IHI [30], the HCBs agreed to use the code 'Halal Supply Chain' for all communication in freight documents, freight labels and ICT systems. As stated by a HCB representative, proper identification is a prerequisite for effective segregation. Furthermore, consensus was achieved on the need for:

1. Clear marking and identification of halal locations and sections in a halal warehouse and halal compliant terminal
2. Well-organized and retrievable Halal Assurance System documentation records
3. Well-organized and retrievable cleaning records (including those related to ritual cleansing) where required.

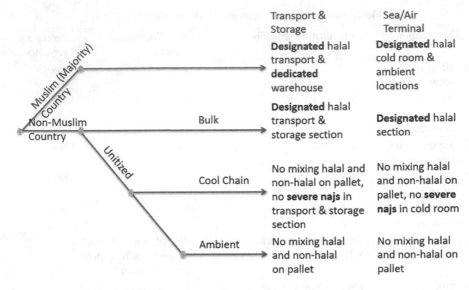

Fig. 1. Segregation in halal supply chains.

5 Conclusion

This paper is part of a large blockchain project being undertaken by a group of researchers at UMP on the application of blockchain technology for halal food, cosmetics, and pharmaceutical supply chains. A large group comprising HCB representatives met in Malaysia in 2017 to discuss halal issue scenarios in halal supply chains; the role of the blockchain system in each halal issue scenario; halal supply chain parameters for the blockchain; and halal supply chain segregation and communication requirements for Muslim and non-Muslim countries.

Halal issues can be categorized into three problem areas, namely contamination, non-compliance and perception. The discussion group agreed to rely on blockchain technology to provide all the related information necessary to manage halal issues when dealing with contamination and non-compliance cases only. (It was agreed that no information would be required from the blockchain system for perception cases).

The key supply chain parameters were identified as the demographic destination country (Muslim (majority) or non-Muslim country), product characteristics (bulk or unitized), and product environment (ambient/cool chain). Based on these parameters a consensus was reached on the need for the segregation of halal supply chains in designated halal transport, storage and halal compliant terminals, in Muslim (majority) countries, whereas for non-Muslim (majority) countries a more lenient approach was considered permissible.

Effective segregation is only possible with effective communication, where 'halal supply chain' needs to be coded on freight documents, freight labels and in ICT systems. Finally, due to the total lack of research on end-to-end cross border halal supply chains, further case studies are recommended for halal supply chains for both the Muslim and non-Muslim (majority) markets.

Acknowledgement. The authors would like to convey a special thank you to Universiti Malaysia Pahang for its financial support through RDU172208 flagship grant for this paper to be presented at the conference.

References

1. Tieman, M.: Halal supply chain certification: the next frontier in halal certification? Islam and Civilis. Renew. (ICR) **9**(2), 233–236 (2018)
2. Tieman, M., Darun, M.R.: Leveraging blockchain technology for halal supply chains. Islam Civilis. Renew. **8**(4), 554–557 (2017)
3. Tieman, M.: Halal risk management: combining robustness and resilience. J. Islamic Mark. **8** (3), 461–475 (2017)
4. Mohamed, Y.H., Abdul Rahim, A.R., Ma'ram, A., Hamza, M.G.: Halal traceability in enhancing halal integrity for food industry in Malaysia – a review. Int. Res. J. Eng. Technol. **3**(3), 68–74 (2016)
5. Bonne, K., Verbeke, W.: Religious values informing halal meat production and the control and delivery of halal credence quality. Agric. Hum. Values **25**(1), 35–47 (2008)
6. Jaafar, H.S., Endut, I.R., Faisol, N., Omar, E.N.: Innovation in logistics services - halal logistics. In: International Symposium on Logistics, Berlin, Germany, pp. 844–851 (2011)
7. Zailani, S., Arrifin, Z., Abd Wahid, N., Othman, R., Fernando, Y.: Halal traceability and halal tracking systems in strengthening halal food supply chain for food industry in Malaysia (a review). J. Food Technol. **8**(3), 74–81 (2010)
8. Mohd Bahrudin, S.S., Illyas, M.I., Desa, M.I.: Tracking and tracing technology for halal product integrity over the supply chain. In: 2011 International Conference on Electrical Engineering and Informatics, Bandung, Indonesia (2011)
9. Zulfakar, M.H.: Australia 's Halal Meat Supply Chain (AHMSC) Operations : Supply Chain Structure, Influencing Factors and Issues. RMIT University (2015)
10. Gartner: Gartner Top 10 Strategic Technology Trends for 2019 (2018). https://www.gartner. com/smarterwithgartner/gartner-top-10-strategic-technology-trends-for-2019/
11. World Economic Forum: These 5 Technologies Have the Potential to Change Global Trade Forever (2018). https://www.weforum.org/agenda/2018/06/from-blockchain-to-mobile-paym ents-these-technologies-will-disrupt-global-trade/
12. IBM: The food industry gets an upgrade with blockchain (2017). https://www.ibm.com/ blogs/blockchain/2017/06/the-food-industry-gets-an-upgrade-with-blockchain/
13. Kamath, R.: Food traceability on blockchain: Walmart's pork and mango pilots with IBM. J. Br. Blockchain Assoc. **1**(1), 1–12 (2018)
14. Larson, K., Grudens-Schuck, N., Allen, B.L.: Can you call it a focus group? Iowa State University Extension, Ames, Iowa (2004). http://www.extensions.iastate.edu/Publications/ PM1969A.pdf. Accessed 15 Jan 2019
15. Jakim. The recognised foreign halal certification bodies & authorities (2017). http://www. halal.gov.my/ckfinder/userfiles/files/cb/CB%20List%20LATEST%20.pdf
16. MUI: List of approved foreign halal certification bodies (2017). http://www.halalmui.org/ images/stories/pdf/LSH/LSHLN-LPPOM%20MUI.pdf
17. Kitzinger, J.: Introducing focus groups. Br. Med. J. **311**(7000), 299–302 (1995)
18. Grudens-Schuck, N., Allen, B.L., Larson, K.: Focus Group Fundamentals. Iowa State University Extension, Ames, Iowa (2004). http://www.extensions.iastate.edu/Publications/ PM1969B.pdf

19. de Ruyter, K.: Focus versus nominal group interviews: a comparative analysis. Mark. Intell. Plann. **14**(6), 44–50 (1996)
20. Walden, G.R.: Focus group interviewing in the library literature: a selective annotated biography 1996–2005. Ref. Serv. Rev. **34**(2), 222–241 (2006)
21. Chambers, D.H., Munoz, A.M.: Focus-group evaluation of nutrition education displays by Hispanic adults who live in the USA. Health Educ. **109**(5), 439–450 (2009)
22. Maxwell, J.A.: Qualitative Research Design: An Interactive Approach. Applied Social Research Method Series, vol. 42. Sage Publications, Thousand Oaks (2005)
23. Prince, M., Davies, M.: Moderator teams: an extension to focus group methodology. Qual. Mark. Res. Int. J. **4**(4), 207–216 (2001)
24. Wall, A.L.: Evaluating an undergraduate unit using a focus group. Qual. Assur. Educ. **9**(1), 23–31 (2001)
25. Heugens, P.P., Van Riel, C., van den Bosch, F.A.: Reputation management capabilities as decision rules. J. Manag. Stud. **41**(8), 1349–1377 (2004)
26. Al-Qaradawi, Y.: The Lawful and the Prohibited in Islam. Islamic Book Trust, Kuala Lumpur (2007)
27. Tieman, M.: The application of Halal in supply chain management: in-depth interviews. J. Islamic Mark. **2**(2), 186–195 (2011)
28. Wilson, J.A.J., Liu, J.: Shaping the Halal into a brand. J. Islamic Mark. **1**(2), 107–123 (2010)
29. Tieman, M., van der Vorst, J.G., Che Ghazali, M.: Principles in halal supply chain management. J. Islamic Mark. **3**(3), 217–243 (2012)
30. IHI Alliance. ICCI-IHI Alliance Halal Standard: Logistics – IHIAS 0100:2010 (first edition), International Halal Integrity Alliance Ltd., Kuala Lumpur (2010)

Maintaining Fog Trust Through Continuous Assessment

Hasan Ali Khattak[1]([⊠]), Muhammad Imran[1], Assad Abbas[1], and Samee U. Khan[2]

[1] Department of Computer Science, COMSATS University Islamabad,
Islamabad 44550, Pakistan
{hasan.alikhattak,mimran,assad.abbas}@comsats.edu.pk
[2] Department of Electrical and Computer Engineering,
North Dakota State University, Fargo, ND 58105, USA
samee.khan@ndsu.edu

Abstract. Cloud computing continues to provide flexible and efficient way for delivery of services, meeting user requirements and challenges of the time. Software, Infrastructures, and Platforms are provided as services in cloud and fog computing in a cost-effective manner. Migration towards fog instigate new aspects of research for security & privacy. Trust is dependent on measures taken for availability, security, and privacy of users' services as well as data in fog as well as sharing of these statistics with stakeholders. Any type of lapses in measures for security & privacy shatter user's trust. In order to provide a trust worthy security and privacy system, we have conducted a thorough survey of existing techniques. A generic model for trustworthiness is proposed in this paper. This model yields a comprehensive component-based architecture of a trust management system to aid fog service providers to preserve users' Trust in a fog computing environment.

Keywords: Fog computing · Security and privacy · Trust management · Trustworthiness

1 Introduction

Internet is constituting as a significant driving force in development of future smart cities. Among major developments in distributed computing the most significant one has been provision of Software as a service (SAAS) to provide online services, Infrastructure as a service (IAAS) for reducing administration and maintenance cost, and Platform as a service (PAAS) to improve the overall provision of the development configurations. Besides availability, privacy as well information security is the basic requirement for a user's Trust [1]. Fog computing has been proposed as a future direction for provision of services and overall service enhancement [2]. Fog Computing otherwise known as edge computing basically leverages edge nodes for hosting the computational powers hence improving the quality of service as well as reducing latency. Key responsibilities of fog service providers are safety, physical security, and reliable availability of computing resources. Trust management performs a key role and

© Springer Nature Switzerland AG 2019
Y. Xia and L.-J. Zhang (Eds.): SERVICES 2019, LNCS 11517, pp. 129–137, 2019.
https://doi.org/10.1007/978-3-030-23381-5_10

effects utilization, services reliability, and infrastructure as well in a fog computing environment [3].

Organizations are stimulating in the direction of fog computing as it is cost effective, manageable for different time critical requirements and relocating their conventional systems on fog. Different fog deployment models such public, private as well as hybrid fog computing architectures along with on-site and off-site fog computing is a popular model for latency prone computing applications [4].

The fundamental characteristics of a secure computing system include confidentiality, integrity, availability, non-repudiation and authentication [5, 6]. Availability of application security systems and standards in the cloud anticipates ease of use to its end users. However, they do not serve as the only feature of a fog computing trust provisioning system.

A trustable Trust Provisioning System ought to anticipate the vital characteristics for inception of trust along with Trustworthiness [7]. A user's trust can be acclaimed by supplying an invariant availability, privacy and security management techniques in fog environment. Smart city application scenario leveraging fog computing are controlled remotely thus providing service providers ease of access and management. Availability can be achieved by surplus resources on these surrogate servers. Similarly, providing legitimate and secure authentication as well as authorization to obscure users is laborious feat in itself [8, 9]. Significant contributions of this work are as mentioned below:

1. Overview of Existing Trust Management Models
2. Requirements for Trust Management in Fog
3. Proposing Generic Trust Management Model
4. Research Issues and Questions for Trustworthiness

The organization of rest of the manuscript is structured in following these notable sections, that are as follows; In Sect. 2, the paper discusses background and basic terminologies of fog as well as cloud computing. Threats and vulnerabilities affecting fog computing are covered in Sect. 3. Related work on Trust and Trustworthiness establishment is covered in Sect. 3.1. Brief discussion of components of fog computing and trust in fog is carried out in Sect. 5.1. Research questions and related work are described in the Sects. 4 and 5, respectively. Proposed framework is concluded in Sect. 6 along with future directions.

2 Background

Delivery of application and softwares through cloud computing is done via Software as a Service (SaaS) for end users [10]. Online compilers, online image manipulation tools, Online Office suits are all famous examples of SaaS. The users, through their authorized credentials are able to connect to the services and use according to their respective agreements. Here the users don't having any kind of authority on fog or cloud authorizations.

Platform as a Service (PaaS) can be described as allocation of an operating system and clusters of programming languages and their development tools to create and deploy customized applications and services. Microsoft Azure and Google Compute

Cloud are some of the significant examples. Similarly, PaaS permits end-users to have command on application design. However, it doesn't provide them the complete authority over the fog or its physical infrastructure [11].

Infrastructure as a Service (IaaS) provides consumers with the straight access to storage, processing, and other similar computing resources, also permitting them to configure these resources and then run operating systems and software on them. Few other notable examples of IaaS are IBM 's Blue cloud and Amazon's elastic compute cloud (Fig. 1).

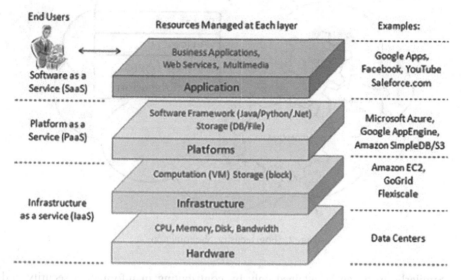

Fig. 1. Fog architectures and services [12]

Fog computing based services are being rendered by a distinct CSPs (Cloud Service Providers). Table 1 contains the list of services and their providers.

Table 1. Cloud services and providers

SaaS	CVM Solutions, Google Apps, Gageln, Host Analytics, Knowledge Tree, Reval, Exoprise Systems, Taleo, NetSuite, Microsoft Office 365, Salesforce.com, Rackspace, Antenna Software, Cloud9 Analytics, IBM and LiveOps
PaaS	Amazon AWS, IBM, Google Apps, Microsoft Azure, Intuit, Netsuite, SAP, SalesForce, WorkXpress, and Joyent
IaaS	OpenStack, Amazon Elastic Compute Cloud EC2, Rackspace, Bluelock, CSC, IBM, Savvis, GoGrid, VMware, Citrix, Terremark, BluePoint and Joyent

3   Fog Computing Security Phenomenon

Fog computing environments contains many components including hardware as well as software which are utilized in order to process huge amounts of data. These can be classified into services or applications, devices and their respective communications as shown in Fig. 2.

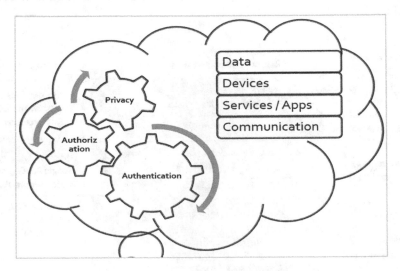

Fig. 2.   Fog security phenomenon

Similarly, trust can be attained only by contributing in information security and privacy expedients, i.e., authorization, authentication and privacy [10, 13]. A trustable trust can be gained by pertaining means for Authorization, Authentication, and Privacy (AAP) at every step and about all the components in the environment. A trustable trust will be delivered even if a person is in a LAN, IoT, WoT, distributed computing and even in fog computing environment [14].

3.1   Trust Concepts

Trust: Trust is an individual's goal for acquiring vulnerability of a trustee, based upon the positive expectancy of their actions.

Trust Establishment: Trust establishment is a procedure and has responsibility of evaluation, maintenance, representation, and distribution of the trust between nodes.

Trust Management: Now, Trust management is presented by Blaze et al. [15] as a consolidated process to specify and interpret security policy guidelines, relationships and their respective credentials that permit uninterrupted authorization of security oriented actions.

Trustworthiness. Trustworthiness on the other hand is an acquired attribute because of the parameters revealed by the respected trustees in a specific environment. Mayer et al. [16], have pointed out 3 significant characteristics which helps us in establishment of the basis for expansion of a trust framework [17]. In the trust development process, integrity, ability, and good will have been highlighted as the key roles of a trustees.

4 Research Questions

Fog computing encourages its stakeholders in many dimensions such as, time and space by achieving reliable connectivity and desirable latency. Flexibility and expandability regarding hardware resources are the main characteristics of fog computing. Efforts and economy of resources are well organized in fog computing. Attraction of users concerning fog computing infrastructure is based on the trust on the fog. The basic research queries to analyze users trust in fog are outlined as follows:

- Measuring Achieved User's trust
- Significance of system feedback on trust management.
- Advantages of sharing statistics affecting user's trust
- Tradeoffs between Trustable trust and Achievable trust
- Tactile Trustworthiness Establishment Matrices
- Feasibility of sharing the information regarding affected users' trust
- Relationship among Trust and Trustworthiness

5 Trust Management System Model

Trustable trust between applications, devices and their users is attained through reinforcing a basic trust management infrastructure. Trust Management Systems (TMS) can be classified into two functions i.e., trustworthiness establishment and most importantly trust establishment of fog modules.

Trust in fog is one of the most important characteristics however, it is also one of the difficult ones and hence, opening doors to many research domains. This paper introduces trust in general and then identifies trustworthiness in fog computing. Diverse cloud computing models are dispensed along with threats and vulnerabilities which each of them may experience. Moreover, a comprehensive list of research questions has been put forward, which results in main subsidy of this paper in form of a proposed Fog Trust Management System model which not only caters those research questions but also sets stone for future research in directions of establishing Trust as well as Trust Management.

Mainly trust in fog is dependent on the user's perception and acceptance of trust and on that trustworthiness of a fog resource can be judged. Trustworthiness of a system is established by means of generating statistics related to trust governance, trust assessment, trust evaluation, event logging and assessment sharing. User Feedback plays a vital role in not only gaining trust but also trustworthiness of any service provided. Trust Assessment sharing enhances the trustworthiness and will enable users to keep track of the services provided and as well as the trust level which service providers are offering. Though cloud service providers are obliged to ensure the availability of these statistics always, otherwise the whole trust phenomenon would be considered incomplete. Next comes the question how much trust will be considered trustable trust, which can be summed up to the top most level of possible achievable trust offered by the system and this is directly measured by the parameters of trust itself namely, Availability, Security and Privacy.

5.1 Trustworthiness

Trustworthiness is to be considered as the ability, benevolence, and therefore integrity of a trustee [18]. The procedures convoluted to inaugurate trustworthiness is outlined as follows:

Trust Governance. A central trust governance characteristic will yield a technique to establish procedures, policies, certification and decertification of artefacts in the cloud [19].

Trust Evaluation. Trust evaluation is the main component for developing trustworthiness of items that are being utilized in fog i.e., software along with hardware and their communication mediums.

Event Logging. The cloud monitoring can be gained successfully by event logging means which are held to manage and govern trust in the fog. Evaluation and recording of fog actions generated events are utilized in developing trustworthiness.

Trust Assessment. Trust assessment is sustained based on the event monitoring and trust evaluation. Dispensing of trust assessment outcomes with the Fog users is the force multiplier in developing trustworthiness.

Psychological Factors. Fog service providers always obtain some good or bad repute from their users. Reputation relies on the satisfaction of users regarding Service Level Agreements (SLAs) [20]. User's trust can easily be transfigured and rectified by these psychological factors.

Assessment Sharing. Finally Trust assessment is dispensed effortlessly through continuous dissemination of statistics on the overall trust situations. Trust assessment outcomes are handed out with the users of the fog, and validation procedure is held based on the feedback of users. When user satisfaction level is attained, and an honorable reputation is developed, trustworthiness is issued to the world (Fig. 3).

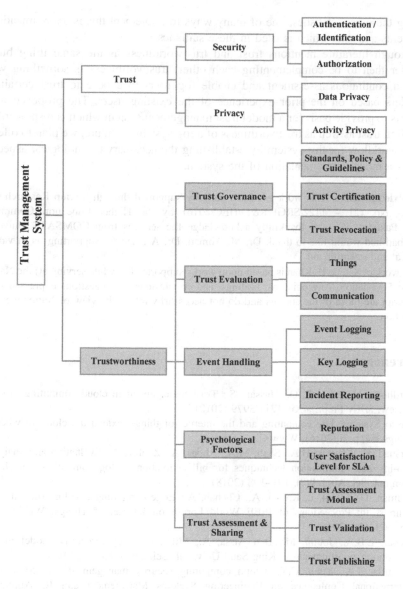

Fig. 3. Basic components of the proposed trust management system for fog

6   Conclusion and Future Directions

The model proposed for Trust Management System emphasizes the total functionality on the trustworthiness subsystem. This module covers all the basic components of the proposed system which are responsible for trust enabled governance, trust establishment, trust assessments, event logging as well as statistics sharing. Fog computing is often on public internet, which makes it essential to offer necessary privacy while sharing these user statistics, one of many ways to implement this is anonymization of the events and data which is used in these statistics.

Though literature mentions trust and trustworthiness as the same thing but we consider then to be complementing each other, trustworthiness is something which needs a continuous assessment and enable fog users to choose to trust certain fog providers based on the prior experience of the existing users. The proposed model attempts to provide basis for a model trust management system which is responsible for providing trust as well as trustworthiness of a fog system. In future, we plan to offer the model workflow for this system by establishing the necessary technological aspects as well as a model implementation of the system.

Acknowledgements. This work has been partially supported through Startup Research Grant Projects No. (21 – 1122/SRGP/R&D/HEC/2016) by the Higher Education Commission (HEC) Pakistan. We also thankfully acknowledge the services from COMSATS University, Islamabad and would like to thank Dr. M. Ahmad, Dr. A Khan for supporting us in valuable technical and scientific aid.

The work of Samee U. Khan is based upon works supported by (while serving at) the National Science Foundation. Any opinions, findings, and conclusions or suggestions expressed in this manuscript are those of the authors and do not necessarily reflect the view of National Science Foundation.

References

1. Firdhous, M., Ghazali, O., Hassan, S.: Trust management in cloud computing: a critical review. arXiv Prepr. arXiv:1211.3979 (2012)
2. Cisco Systems. Fog computing and the internet of things: extend the cloud to where the things are, p. 6 (2016). Www.Cisco.Com
3. Arshad, H., Khattak, H.A., Shah, M.A., Abbas, A., Zoobia, A.: Evaluation and analysis of bio-inspired optimization techniques for bill estimation in fog computing. Int. J. Adv. Comput. Sci. Appl. 9(7), 191–198 (2018)
4. Salman, O., Elhajj, I., Kayssi, A., Chehab, A.: Edge computing enabling the Internet of Things. In: Proceedings of IEEE World Forum on Internet of Things, WF-IoT 2015, pp. 603–608 (2016)
5. Rabai, L.B.A., Jouini, M., Ben Aissa, A., Mili, A.: A cybersecurity model in cloud computing environments. J. King Saud Univ. Inf. Sci. 25(1), 63–75 (2013)
6. Almulla, S.A., Yeun, C.Y.: Cloud computing security management. In: 2010 Second International Conference on Engineering Systems Management and Its Applications (ICESMA), pp. 1–7 (2010)
7. Nitti, M., Girau, R., Atzori, L., Iera, A., Morabito, G.: A subjective model for trustworthiness evaluation in the social internet of things. In: 2012 IEEE 23rd International Symposium on Personal, Indoor and Mobile Radio Communications-(PIMRC), pp. 18–23 (2012)
8. Martucci, L.A., Zuccato, A., Smeets, B., Habib, S.M., Johansson, T., Shahmehri, N.: Privacy, security and trust in cloud computing: the perspective of the telecommunication industry. In: 2012 9th International Conference on Ubiquitous Intelligence & Computing and 9th International Conference on Autonomic & Trusted Computing (UIC/ATC), pp. 627–632 (2012)

9. Li, X., Du, J.: Adaptive and attribute-based trust model for service level agreement guarantee in cloud computing. IET Inf. Secur. **7**(1), 39–50 (2013)
10. Stojmenovic, I., Wen, S., Huang, X., Luan, H.: An overview of fog computing and its security issues. Concurrency Comput. Pract. Exp. **28**(10), 2991–3005 (2015)
11. Liang, G.: Automatic traffic accident detection based on the internet of things and support vector machine. Int. J. Smart Home **9**(4), 97–106 (2015)
12. Stojmenovic, I., Wen, S.: The fog computing paradigm: scenarios and security issues. In: Proceedings of 2014 Federated Conference on Computer Science and Information Systems, vol. 2, pp. 1–8 (2014)
13. Khan, S.U., Khan, R.: Content-location based key management scheme for content centric networks. In: Proceedings of 6th International Conference on Security of Information and Networks - SIN 2013, pp. 376–379 (2013)
14. Muzammal, S.M., et al.: Counter measuring conceivable security threats on smart healthcare devices. IEEE Access **6**, 20722–20733 (2018)
15. Blaze, M., Feigenbaum, J., Ioannidis, J., Keromytis, A.D.: The role of trust management in distributed systems security. In: Vitek, J., Jensen, C.D. (eds.) Secure Internet Programming. LNCS, vol. 1603, pp. 185–210. Springer, Heidelberg (1999). https://doi.org/10.1007/3-540-48749-2_8
16. Mayer, R.C., Davis, J.H., Schoorman, F.D.: An integrative model of organizational trust. Acad. Manag. Rev. **20**(3), 709–734 (1995)
17. Noor, T.H., Sheng, Q.Z.: Trust as a service: a framework for trust management in cloud environments. In: Bouguettaya, A., Hauswirth, M., Liu, L. (eds.) WISE 2011. LNCS, vol. 6997, pp. 314–321. Springer, Heidelberg (2011). https://doi.org/10.1007/978-3-642-24434-6_27
18. Colquitt, J.A., Scott, B.A., LePine, J.A.: Trust, trustworthiness, and trust propensity: a meta-analytic test of their unique relationships with risk taking and job performance. J. Appl. Psychol. **92**(4), 909 (2007)
19. Kang, D., Jung, J., Mun, J., Lee, D., Choi, Y., Won, D.: Efficient and robust user authentication scheme that achieve user anonymity with a Markov chain. Secur. Commun. Netw. **9**(11), 1462–1476 (2016)
20. Aazam, M., St-Hilaire, M., Lung, C.-H., Lambadaris, I., Huh, E.-N.: IoT resource estimation challenges and modeling in fog. In: Rahmani, A.M., Liljeberg, P., Preden, J.-S., Jantsch, A. (eds.) Fog Computing in the Internet of Things, pp. 17–31. Springer, Cham (2018). https://doi.org/10.1007/978-3-319-57639-8_2

Study on the Coordination Degree Between FDI and Modern Service Industry Development in Shenzhen

Xiangbo Zhu[1,2($\boxtimes$)], Guoliang Ou[1], and Gang Wu[1]

[1] Shenzhen Polytechnic,
Shenzhen 518055, Guangdong, People's Republic of China
zxb@szpt.edu.cn
[2] China Three Gorges University,
Yichang 443002, Hubei, People's Republic of China

Abstract. Foreign direct investment (FDI) is an important driving force for economic development. In the period of rapid development and growth of Modern Service Industry (MSI), whether FDI had any impact on it was not clear. Through selecting some indicators form scale, structure and performance, this paper constructs the "FDI-MSI" coordination degree evaluate model. Based on the data of foreign direct investment (FDI) and modern service industry (MSI) collected from Shenzhen, the development coordination degree between foreign direct investment (FDI) and modern service industry (MSI) was estimated in 2010–2018. Research results show that the foreign direct investment and the modern service industry are generally coordinated in Shenzhen, but the overall level is not high. The structure of foreign capital is not reasonable, and the current inefficiency in the use of foreign capital by the service industry is the main reason. In the process of utilizing foreign capital in the future, Shenzhen should strengthen the guidance of foreign investment with the goal of modern service industry development, create a favorable environment for foreign capital to flow into modern service industry, and promote the coordinated development of foreign capital utilization and modern service industry.

Keywords: Shenzhen · FDI · Modern service industry · Coordination degree

1 Introduction

With the deepening of service economics, the role of modern service industry in economic development is increasing. In 2015, the added value of Shenzhen's modern service industry reached 713.4 billion yuan, accounting for 69.3% of the added value of the service industry. Cross-border finance, factor market, wealth management and other innovative financial fields have developed rapidly. The traditional logistics industry has gradually transformed into supply chain management and e-commerce logistics. Culture and technology, tourism and commerce are highly integrated, and high-end features are initially revealed. High-tech content & high value-added service industry has developed rapidly. However, Shenzhen's modern service industry still has many

© Springer Nature Switzerland AG 2019
Y. Xia and L.-J. Zhang (Eds.): SERVICES 2019, LNCS 11517, pp. 138–147, 2019.
https://doi.org/10.1007/978-3-030-23381-5_11

problems in terms of scale, regional coordination, support system, and external expansion.

Internationalization is an important force in the development of modern service industry. Both theory and practice have proved that foreign direct investment (FDI) has a significant role in promoting the development of modern service industry. As an important city for attracting and utilizing foreign capital in China, Shenzhen can effectively promote the smooth realization of economic development goals by continuously raising the level and level of modern service industry by utilizing foreign capital. It is necessary to explore the relationship between the use of foreign capital and the development of modern service industry in Shenzhen, with the using of coordination degree.

2 Literature Review

For foreign direct investment and modern service industry, domestic and foreign scholars have carried out research from different angles and have obtained rich and varied research results.

The birth and development of modern service industry. Shen (2012) comprehensively introduced the composition and characteristics, development environment and basic conditions of Shenzhen's modern service industry. Applying the proportion of the added value of modern service industry, the contribution rate of growth, the driving force of economic growth, etc., analyze the development goals and the current situation and future path of the modern service industry. Cao (2010) used the Malmquist index method to measure the development efficiency of the modern service industry in the Pearl River Delta region, and conducted a comparative analysis of the modern service industry development model in nine cities in the region. The results show that the growth model of the modern service industry in the Pearl River Delta is no longer It is an extensive type in the past, and the improvement of the efficiency of factor use has accelerated the development of the modern service industry. Wang (2013) believes that the development of modern service industry is inseparable from financial support, and the role of financial capital allocation plays an important role in promoting the development of modern service industry. Cui and Li (2011) proposed that the development of modern service industry is dynamic, and its development and evolution has a systematic dynamic mechanism. Ma and Li (2016) constructed Service Industry adjustment range degree index from four dimensions, including Structure Change Service Index, Service Industry Structure Rationalization Index, Service Industry Structure Advanced Index, and Service Industry Structure, and inferred adjustment rationality, adjustment direction and adjustment path, and empirically measured the structural adjustment of China's service industry from 1992 to 2012. Deng et al. (2012) established China's modern service industry evaluation index system, which includes four aspects: basic conditions, development level, specialization degree and growth potential. Huang (2016) believes that big data service industry is the general trend of modern service industry. He discusses the status quo, application value and industrial structure of big data service industry in China, and the technology existing in the development of big data service industry. Issues such as privacy protection, application methods and development strategies. Yang (2014) believes that it is necessary to solve the human resources problems of modern service industry, such as insufficient talents, low quality and unreasonable structure, etc. The key point is to build an integrated talent training mechanism. Based on the empirical research, Qiao (2013) analyze the key issues that constrain the rapid development of the industry, and to examine the policy experience of leading countries and regions in the world, and put forward proposals to accelerate the development of China's modern service industry.

The relationship between modern service industry and foreign direct investment. Blind and Jungmittag (2014) argue that foreign direct investment (FDI) has a significant role in promoting product and service innovation for service companies. An empirical study by Brooks (2008) shows that there is a one-way causal relationship between FDI levels and service exports in the short term. Chakraborty and Nunnenkamp (2006) pointed out that FDI has a short-term effect on the output of the service industry. Francois and Woerz (2007) concluded that increased openness in services with exports and FDI has a positive effect on exports, value added and manufacturing employment. The measurement results of Fernandes and Paunov (2008) found that service industry FDI can promote manufacturing productivity. Doytch and Uctum (2016) found that, except for low- and middle-income countries, manufacturing FDI has no significant effect on service sector growth, while service industry FDI has significantly contributed to service sector growth.

In general, relevant research at home and abroad has been relatively mature, but there are still some places for further research. First of all, the selection of indicators in the existing research is relatively single, failing to select comprehensive and systematic indicators, and cannot reflect the comprehensive effect of FDI on the development of the service industry. At the same time, research samples are mostly in the country, and research on a certain city or region is rare. Secondly, the existing research mostly adopts methods such as cointegration and Granger causality test. These studies have high requirements on data volume and are not suitable for city level. Based on the above analysis, this paper will try to construct the "FDI-MSI" composite system coordination degree model based on the synergistic thinking, apply the FDI and MSI data of Shenzhen during 2009–2017, and measure the coordination between them. It is expected to explore the development of Shenzhen's foreign direct investment and modern service industry, and provide theoretical and empirical support for Shenzhen to better utilize foreign capital to promote the development of modern service industry.

3   Composite System Coordination Degree Model Construction

The coordinated development refers to a virtuous circle formed by the cooperation, mutual cooperation and mutual promotion between the FDI subsystem and the MSI subsystem and the components of each subsystem in the composite system of "FDI-MSI". The situation, the synergy between the subsystems determines the evolution of the system from disorder to order. The key mechanism of the system from disorder to order is the synergy between the internal order parameters of the system. It influences the characteristics and laws of the system phase change. The degree of coordination is a measure reflecting this synergy. This paper draws on the research results of Yafei et al. (2012) to construct the coordination model of Shenzhen "FDI-MSI" composite system.

3.1 Establish of Efficiency Function

The contribution of order parameters to system order is usually expressed by the efficiency coefficient (EC), and the relationship describing EC is called the efficiency function. If V_{ij} refers to the order parameters, than $EC(V_{ij}) = F(V_{ij})$, which F means the Functional relationship of efficiency, $i(i = 1, 2)$ is the subscript of subsystem, $j(j = 1, 2, \ldots, n)$ is the subscript of subsystem order parameters. The result value of order parameters V_{ij} is $X_{ij}(i = 1, 2; j = 1, 2, \ldots, n)$, while α_{ij} and β_{ij} refer to the critical point upper limit and lower limit of order parameters V_{ij} when system is stable, that means $\beta_{ij} \leq X_{ij} \leq \alpha_{ij}$.

When the system state is stable, the change of order parameters had two efficiency function on the ordered degree of system. One is positive, that is, as the variable increases, the order of the system increases. One is the negative effect, that is, as the variable increases, the order of the system decreases. Efficiency function $EC(V_{ij})$ of order parameters V_{ij} can expressed as follows:

$$\text{If } EC(V_{ij}) \text{ is positive effect, } EC(V_{ij}) = \frac{X_{ij} - \beta_{ij}}{\alpha_{ij} - \beta_{ij}}, \ \beta_{ij} \leq X_{ij} \leq \alpha_{ij} \tag{1}$$

$$\text{If } EC(V_{ij}) \text{ is negative effect, } EC(V_{ij}) = \frac{\alpha_{ij} - X_{ij}}{\alpha_{ij} - \beta_{ij}}, \ \beta_{ij} \leq X_{ij} \leq \alpha_{ij} \tag{2}$$

3.2 Establish of System Ordered Degree Function

In addition, used the geometric mean of efficiency function to calculate the system ordered function $EC_i(V_i)$, which is:

$$EC_i(V_i) = \sqrt[n]{\prod_{j=1}^{n} EC_i(V_{ij})}, \ i = 1, 2; \ j = 1, 2, \ldots, n \tag{3}$$

For $EC(V_{ij}) \in [0, 1]$, so the value of $EC_i(V_i)$ is also between 0 and 1. When $EC_i(V_i)$ equals to 0, that means the system ordered is the minimum. When $EC_i(V_i)$ equals to 1, that means the system ordered is the maximum.

3.3 Definition of Composite System Coordination Degree

For a given moment t_0, the ordered degree of all the subsystem is $EC_i^0(V_i)$. While to the any random moment t_1 during evolution process of the composite system, the ordered degree of all the subsystem is $EC_i^1(V_i)$. Based on that, the coordination degree cm of composite system can be definite as follows:

$$cm = \theta \sqrt{\left| \prod_{i=1}^{2} [EC_i^1(V_i) - EC_i^0(V_i)] \right|}, \ i = 1, 2, \ \text{while,} \tag{4}$$

$$\theta = \frac{\min_i [EC_i^1(V_i) - EC_i^0(V_i)]}{\left| \min_i [EC_i^1(V_i) - EC_i^0(V_i)] \right|}, \ i = 1, 2 \tag{5}$$

For cm and θ, we make the following instructions: ① cm should include all subsystems and $cm \in [-1, 1]$. As bigger the cm is, the higher coordination degree of composite system, and vice versa. ② If and only if $EC_i^1(V_i) - EC_i^0(V_i) > 0, \theta$ and coordination degree of composite system has the positive value.

4 Empirical Analysis of Development Coordination Degree on the "FDI-MSI" Composite System in Shenzhen

4.1 Establish Index System

The "FDI-MSI" system includes an FDI subsystem and an MSI subsystem. From the three aspects, such as scale, structure and performance, to construct the evaluation index system of Shenzhen "FDI-MSI" composite system coordination degree.

For the foreign direct investment (FDI) subsystem, the actual use of foreign capital is used to characterize the FDI scale; the actual use of foreign services in the modern service industry accounts for the proportion of actual use of foreign investment to measure the FDI structure; the FDI performance index is used to measure FDI performance, of which performance The index is calculated using the following formula:

$$FDI \ performance \ index = \frac{FDI \ Scale \ of \ shenzhen/FDI \ Scale \ of \ Guangdong}{GDP \ Scale \ of \ shenzhen/GDP \ Scale \ of \ Guangdong}$$

For the modern service industry subsystem, the modern service industry GDP is used to measure the scale of modern service industry output, the number of modern service industry employees is used to measure the scale of modern service industry employment; the modern service industry labor productivity is used to measure the performance of modern service industry, through the modern service industry GDP. Calculated from the number of employees in the modern service industry.

4.2 Empirical Data Collection

In view of foreign direct investment, in accordance with the policy provisions of the "Guidance on Foreign Investment Directions" and the "Guidance Catalogue for Foreign Investment Industries" promulgated by the Chinese government, the industry division of foreign direct investment is clarified. In conjunction with the statistics of foreign direct investment by the National Bureau of Statistics and the Shenzhen Municipal Bureau of Statistics, relevant data were obtained from the Shenzhen

Statistical Yearbook (2005–2018). According to the National Development and Reform Commission's "Industrial Structure Adjustment Guidance Catalogue" (2019, Exposure Draft), the modern service industry includes the following nine major industries: (1) information transmission, computer services and software industry; (2) financial industry (3) real estate; (4) leasing and business services; (5) scientific research, technical services and geological exploration; (6) water conservancy, environment and public facilities management; (7) education; (8) Health, social security and social services; (9) culture, sports and entertainment. Through the "Shenzhen Statistical Yearbook" (2010–2018) to obtain relevant data, and calculate it. The raw data of the coordination evaluation of the "FDI-MSI" composite system in Shenzhen is shown in Table 1:

Table 1. Raw data of the coordination evaluation of the "FDI-MSI" composite system in Shenzhen

Year	FDI subsystem			MSI subsystem			
	FDI scale	FDI structure	FDI performance	MSI output scale	MSI labor scale	MSI structure	MSI performance
2004	2.35	0.112611	2.1732	141.7741	75.5909	0.683692	18.75544
2005	2.969	0.098726	2.8756	158.9147	88.1686	0.686058	18.02395
2006	3.269	0.12789	2.4351	1931104	100.3293	0.69745	19.24766
2007	3.662	0.157247	2.7328	246.4986	121.5746	0.724149	20.2755
2008	4.030	0.345801	2.1334	270.5695	124.4807	0.684436	21.73586
2009	4.160	0.290761	2.9824	311.7353	126.7308	0.697505	24.59823
2010	4.297	0.394667	2.6103	369.1031	141.6156	0.702063	26.06373
2011	4.599	0.377797	2.7558	436.6289	149.4667	0.705671	29.21245
2012	5.229	0.357782	2.7827	510.0661	159.6889	0.702865	31.94124
2013	5.468	0.502144	2.4133	574.0036	182.8385	0.690314	31.39402
2014	5.805	0.594113	2.9526	629.9816	185.7097	0.683316	33.92292
2015	6.497	0.756111	2.6654	696.84	196.0708	0.67466	35.54022
2016	6.732	0.824509	2.3489	774.9147	207.5934	0.658898	37.32848
2017	7.401	0.705153	2.5879	828.1394	222.6925	0.629649	37.18757

Note: The unit of FDI Scale is Billion U.S. dollars; The unit of MSI Output Scale is Billion RMB; The unit of MSI Labor Scale is Ten thousand people.

4.3 Calculation Results

First, the efficiency coefficient of each indicator is calculated according to formula (1) and formula (2). According to the research of Yanping and Mingsheng (2011), we set the upper limit and the lower limit of the indicator. While the upper limit of the indicator is set to the 110% of the maximum value, and the lower limit of the indicator is set to the 90% of minimum value. According to this rule, the power efficiency coefficients of each indicator are calculated as shown in Table 2 below:

Table 2. Indicators power efficiency coefficients of "FDI-MSI" composite system in Shenzhen

Year	FDI subsystem			MSI subsystem			
	FDI scale	FDI structure	FDI performance	MSI output scale	MSI labor scale	MSI structure	MSI performance
2004	0.039173	0.02904	0.236554	0.018097	0.042724	0.508996	0.110782
2005	0.142355	0.012068	0.720774	0.039976	0.113812	0.519288	0.078801
2006	0.192362	0.047716	0.417102	0.083626	0.182544	0.568845	0.132302
2007	0.257872	0.0836	0.62233	0.151774	0.302621	0.684988	0.17724
2008	0.319215	0.314076	0.140178	0.182499	0.319047	0.512232	0.241087
2009	0.340884	0.246799	0.794399	0.235046	0.331764	0.569084	0.36623
2010	0.363721	0.373807	0.537881	0.308273	0.415892	0.588912	0.430303
2011	0.414062	0.353186	0.638186	0.394468	0.460266	0.604607	0.567965
2012	0.519078	0.328721	0.65673	0.488207	0.518042	0.5924	0.687269
2013	0.558917	0.505179	0.402074	0.569821	0.648882	0.537802	0.663344
2014	0.615092	0.617596	0.773856	0.641275	0.66511	0.50736	0.773908
2015	0.730443	0.815612	0.575866	0.726617	0.723671	0.469706	0.844617
2016	0.769615	0.899217	0.357678	0.826276	0.788796	0.40114	0.9228
2017	0.881132	0.753324	0.522439	0.894215	0.874135	0.273904	0.91664

Secondly, according to formula (3), calculate the ordered degree of FDI subsystem and the ordered degree of MSI subsystem, and calculate the coordination degree of Shenzhen "FDI-MSI" composite system according to formulas (4) and (5). The calculation results are as shown in the following Table 3:

Table 3. Ordered degree of subsystem and the coordination degree of composite system

Year	Ordered degree of subsystem		Coordination degree of composite system	
	FDI subsystem	MSI subsystem	Status	Coordination degree
2004	0.064561	0.081258	0	–
2005	0.107382	0.116811	1	0.018816
2006	0.156438	0.184106	1	0.087039
2007	0.237617	0.273266	1	0.17132
2008	0.241323	0.291198	1	0.18109
2009	0.405816	0.357049	1	0.280418
2010	0.418184	0.424556	1	0.331127
2011	0.453599	0.499694	1	0.400961
2012	0.482113	0.566471	1	0.455352
2013	0.484209	0.602652	1	0.458444
2014	0.664918	0.639714	1	0.520595
2015	0.700052	0.675823	1	0.544475
2016	0.62788	0.700847	1	0.534446
2017	0.702565	0.665586	1	0.487665

It can be seen from the calculation results of Tables 2 and 3 that the subsystem and system coordination degree of Shenzhen "FDI-MSI" composite system has the following characteristics:

First, both the foreign direct investment subsystem and the modern service industry subsystem are moving in an orderly direction. It shows the foreign direct investment amount and the development of modern service industry in Shenzhen, all showing the trend of scale growth, structural optimization and performance improvement. From the perspective of the FDI subsystem, the ordering coefficient has increased from 0.064561 in 2004 to 0.702565 in 2017. The growth trend is obvious, but it has declined in 2016. From the perspective of the sub-index "FDI performance", the value in 2016 has been significantly reduced, directly leading to a decline in the ordering coefficient. Since 2016, the international economic situation has continued to deteriorate, directly affecting the level and performance of Shenzhen's actual use of foreign capital. From the perspective of the MSI subsystem, the ordering coefficient continued to increase in 2004–2017, reaching a maximum of 0.700847 in 2016 and falling back to 0.665586 in 2017.

Secondly, from the comparison of foreign investment subsystem and modern service industry subsystem, the order of modern service industry subsystem is higher than that of foreign direct investment subsystem in 2004–2008, indicating the development of foreign direct investment to modern service industry. The promotion needs to be improved, and the rapid development of the modern service industry has attracted both foreign investment and the efficient use of foreign capital. In 2009–2017, the order of foreign direct investment subsystems was higher than that of modern service industry subsystems, indicating that the use of foreign capital promoted the orderly development of modern service industry to a certain extent and improved the development quality of modern service industry.

Thirdly, from the perspective of the coordination degree of the composite system, with the year of 2004 as the base period, the coordination degree of the "FDI-MSI" composite system in Shenzhen is on the rise during 2005–2017, reaching a maximum of 0.534446 in 2016. In terms of absolute value, the value of coordination in 2005 was 0.087039, and the value of coordination in 2017 was 0.487665, which was a significant increase. From the relative value point of view, the coordination degree of the composite system increased rapidly in 2004–2011, and the speed of coordination improvement in 2012–2017 slowed down, but the momentum was stable.

5   Conclusion

The use of foreign capital in Shenzhen has been continuously improved in scale, structure and performance, showing a situation of coordinated development with the modern service industry. However, the problems of unreasonable foreign capital structure and foreign investment performance index in the process of using foreign capital in Shenzhen have affected the role of foreign capital in promoting the development of the service industry. From the results of the composite systemic coordination degree, although the level of coordination continues to rise, it is still in a lower state. In 2015, the level of coordination reached the highest level, which was only 0.544475.

In theory, using foreign capital to increase the supply capacity of modern service industry and optimize the structure of service industry can continuously improve the level and level of modern service industry and promote the smooth realization of its economic development goals. As a special economic zone, Shenzhen has unique and powerful conditions in terms of preferential policies, geographical location, talent reserves, transportation facilities, industrial agglomeration and technological innovation. These advantages have attracted more and more foreign investment. The large-scale and efficient use of foreign capital has become an important reason for the rapid development of Shenzhen.

In the future practice, Shenzhen should aim at the rapid and high-quality development of modern service industry and guide foreign capital to better flow into the modern service industry. Accelerate the construction of a new pattern of service industry development that is more innovative and dynamic, highlighting the characteristics of modern and mega-city, and promote the service industry to broaden the field, enhance functions, and optimize the structure. By relaxing the restrictions on foreign investment in modern service industries, we will formulate policies that are conducive to the flow of foreign capital into modern service industries, and form a good situation in which foreign capital utilization and modern service industries are highly coordinated.

Acknowledgements. This research was supported by Humanities and Social Sciences Youth Project of Hubei Provincial Department of Education (No. 17Q065), Humanities and Social Sciences Annual Project of Shenzhen Polytechnic (No. 602930239), Soft Science research project of Technology Bureau of Yichang (No. A18-30305), Soft Science research project of Hubei Provincial Department of Education (No. 2018ADC162), Open fund project of Hubei Provincial Humanities and Social Sciences Key Research Base– Reservoir Immigration Research Center (No. 2016KF11).

References

Shen, X., Tan, L., Xiao, C., Gong, J.: Report of Shenzhen economic development, pp. 177–187. Social Sciences Academic Press, Beijing (2012)

Cao, J.: Study on total factor productivity of modern service industry of Pearl River Delta. J. Gansu Adm. Inst. **2**, 91–97 (2010)

Wang, S.: Analysis of the effect of financial support for the development of suzhou modern service industry. Bus. Econ. **22**, 112–114 (2013)

Ma, F., Li, J.: Service industrial structure change in china: measures based on four dimensions. Econ. Manag. J. **2**, 26–33 (2016)

Deng, Z., Hu, S., Zhang, W.: Evaluation index system of modern service industry and its empirical analysis. Technol. Econ. **31–10**, 60–63 (2012)

Huang, H.: Big data service industry: trends, challenges and countermeasures. Res. Dev. **3**, 16–20 (2016)

Yang, L.: Research on talents training strategy of modern service industry under the background of China's economic transition. Reform. Strategy **4**, 127–131 (2014)

Qiao, W.: Research on policy issues of modern service industry–empirical research and international experience, pp. 206–220. Social Sciences Academic Press, Beijing (2013)

Blind, K., Jungmittag, A.: Foreign direct investment, imports and innovations in the service industry. Rev. Ind. Organ. **25**, 205–227 (2014)

Brooks, D.H., Roland-Holst, D., Zhai, F.: Behavioral and empirical perspectives on FDI: International capital allocation across Asia. J. Asian Econ. **19**, 40–52 (2008)

Fernandes, A.M., Paunov, C.: Foreign direct investment in services and manufacturing productivity growth. The World Bank Policy Research Working Paper, WPS4730 (2008)

Chakraborty, C., Nunnenkamp, P.: Economic Reforms, FDI and its Economic Effects in India. Working Paper No. 1272. The Kiel Institute of World economy, Germany (2006)

Francois, J., Woerz, J.: Producer Services, Manufacturing Linkage, and Trade. Tinbergen Institute Discussion Paper, No. 045/2 (2007)

Doytch, N., Uctum, M.: Globalization and the environmental impact of sectoral FDI. Econ. Syst. **40**(4), 582–594 (2016). Elsevier

Yafei, L., Haifeng, W., Xiaoyang, F.: International comparison on the degree of coordination between economic development and HERD investment. Sci. Res. Manag. **33**(4) (2012). Elsevier

Yanping, Z., Mingsheng, L.: Evaluation on coordination and matching degree between regional talent structure optimization and industrial structure upgrade in China, no. 3 (2011)

Author Index

Printed in the United States
By Bookmasters